KENTUCKY REMEMBERED
An Oral History Series

James C. Klotter
and
Terry Birdwhistell
GENERAL EDITORS

Tobacco Culture

Farming Kentucky's Burley Belt

JOHN VAN WILLIGEN
SUSAN C. EASTWOOD

THE UNIVERSITY PRESS OF KENTUCKY

Copyright © 1998 by The University Press of Kentucky

Scholarly publisher for the Commonwealth,
serving Bellarmine College, Berea College, Centre
College of Kentucky, Eastern Kentucky University,
The Filson Club Historical Society, Georgetown College,
Kentucky Historical Society, Kentucky State University,
Morehead State University, Murray State University,
Northern Kentucky University, Transylvania University,
University of Kentucky, University of Louisville,
and Western Kentucky University.
All rights reserved

Editorial and Sales Offices: The University Press of Kentucky
663 South Limestone Street, Lexington, Kentucky 40508-4008

02 01 00 99 98   1 2 3 4 5

Library of Congress Cataloging-in-Publication Data
Van Willigen, John
  Tobacco culture : farming Kentucky's burley belt / John van Willigen, Susan C. Eastwood.
    p. cm.
  Includes bibliographical references and index.
  ISBN 0-8131-2066-7 (alk. paper)
  1. Burley tobacco—Kentucky—History. I. Eastwood, Susan C., 1954- . II. Title.
SB273.V35   1998
338.1'7371'09769—dc21   97-46321

This book is printed on acid-free recycled paper meeting
the requirements of the American National Standard
for Permanence of Paper for Printed Library Materials.

Manufactured in the United States of America

Contents

General Editors' Preface vii

Preface viii

1. Tobacco Culture 1

2. Tobacco Ground 14

3. Tobacco Labor 25

4. The Tobacco Program 38

5. Sowing the Beds 64

6. Setting the Plants 83

7. Cultivating and Topping 102

8. Cutting, Housing, and Curing 116

9. The Stripping Room 142

10. On the Floor 161

11. Burley Tobacco and Its Transformations 179

Notes 194

References Cited 196

Index 201

General Editors' Preface

In the field of oral history, Kentucky is a national leader. Over the past several decades, thousands of its citizens have been interviewed. *Kentucky Remembered* brings into print the most important of those collections, with each volume focusing on a particular subject.

Oral history is, of course, only one type of source material. Yet by the very personal nature of recollection, hidden aspects of history are often disclosed. Oral sources provide a vital thread in the rich fabric that is Kentucky history.

This volume is the fourth in the series and focuses on one of the most important parts of Kentucky's past and present—tobacco culture. Kentucky's roots are deeply imbedded in the process of tobacco growing. Through the voices of Kentucky's tobacco farmers, *Tobacco Culture* explores the process of growing tobacco and the culture that has developed around that process.

Even as the national debate regarding tobacco rages throughout the country, John van Willigen and Susan Eastwood offer insight into the lives of those whose livelihood depends in large part on tobacco production. Oral history is at its best when it provides a forum for those most often ignored. *Tobacco Culture* helps provide that forum.

Preface

This work concerns the cultural world of tobacco farming. In it, we describe the knowledge and practices of the men and women of central Kentucky involved in producing white burley tobacco. Through personal experiences, we developed great respect for tobacco people and their craft. We were struck by the ironic contrast between what we saw as vilification of tobacco and tobacco producers in the media and the grace and dignity of these hardworking craftsmen. In effect, this book is a response to the assault on tobacco that may help people remember the cultural world antismoking proponents are attempting to dismantle. Tobacco people and their knowledge are valuable American cultural and economic resources.

Tobacco production takes place in a large, highly contested arena of politics, law, commerce, medical research, and consumer behavior. So, although this book focuses on production and marketing practices, to understand "tobacco culture," we need to keep in mind the increase in concern about the negative health effects of tobacco use and the subsequent political transformations this causes. The cultural meaning of burley tobacco farming has slowly changed since 1964 when the surgeon general's report linking tobacco use with various health problems was published. This change in meaning is the biggest challenge facing tobacco people.

The narratives from which this book was formed are based on interviews with farmers and other people involved in tobacco production and processing in various counties in the heart of the "burley belt" in central Kentucky. Men and women from Bourbon, Fayette, Madison, Mason, Montgomery, Robertson, Rockcastle, and Scott counties were asked to narrate their experiences. These interviews, conducted from 1978 to 1992, document current practices and historic change.

Preface

This book represents a complex collaboration of cultural anthropologists and oral historians and grew out of separate projects conducted by each author. John van Willigen's project documented a broad spectrum of social and economic life in a rural central Kentucky county to provide a context for planning and interpreting a social survey on the social relations of older people (van Willigen 1989). Tobacco production happens to be an important aspect of the social and economic life of the county where that research was done and, therefore, was documented extensively. Susan C. Eastwood's involvement began with a linguistic anthropology graduate research paper on the language of tobacco. This work was used as the basis for a thesis in cultural anthropology (Eastwood 1989). A number of Eastwood's relatives involved in tobacco farming in central Kentucky became participating narrators.

Using materials gained from our separate projects, we began the process of compiling a single text. This received impetus from the initiation of the Kentucky Family Farm Oral History Project (1990-94). Funded by the Kentucky Oral History Commission, this project allowed additional individuals to be interviewed by Andrea Allen and Sonja Anglin. We thank Kim Lady Smith, director of the Kentucky Oral History Commission, for her help with and encouragement of this project.

Tapes from the Oral History Project at the University of Kentucky library and the Commonwealth of Kentucky Oral History Commission also were integrated into the project. The breadth of data on tobacco production was increased further through interviews with farmers and others carried out by John Klee of Maysville Community College, who has an extensive knowledge of the tobacco production process. Deposited at the Kentucky Historical Society and available to the public, they provided a useful supplement to the interviews already obtained. A series of taped monologues by Paul Carraco, a farmer and warehouse owner from Carrollton, Kentucky, was also useful. Use of these tapes was facilitated by Jeffery Suchanek and Terry Birdwhistell of the Margaret I. King Library at the University of Kentucky.

In addition to the interview data, we made use of participant-observation; that is, we performed most steps in the tobacco production process to one extent or another within the limits of our skill, endurance, and willingness to accept danger. We pulled plants, dropped plants, burned seedbeds, topped, cut tobacco, sprayed sucker control chemicals, took crops to market, stripped and tied tobacco, and baled tobacco. Often interviews were done while working, with tape recorder resting on the stripping room table. The interview and observational data are supplemented by print materials, including newspapers and books.

The narrative texts are derived from transcriptions of tape recordings of narrators. The transcriptions were done either by highly experienced professional transcribers or by the authors. Initial transcriptions aimed to yield a verbatim account and were then edited for readability. False starts, hesitations, and repetitions are removed, and words are added in brackets or reordered for clarity. The original tapes and transcripts are available at the University of Kentucky or the Kentucky Historical Society where they can be reviewed.

Earlier drafts were read and commented upon by Terry L. Birdwhistell, C. Milton Coughenour, Billie R. DeWalt, George A. Duncan, James Klotter, Gary K. Palmer, Gil Rosenberg, William M. Snell, Jeffery S. Suchanek, Juliana M. Van Willigen, Anne G. Van Willigen and anonymous readers. Each reader provided invaluable help in clarifying the text as we worked with it.

This book focuses on the production of a single commodity. Readings that contextualize burley tobacco production also may be useful. Virgil Steed's *Kentucky Tobacco Patch* (1947) provides a rich narrative description of the annual cycle of farming activities on his Fayette County farm in the early 1940s. Steed's book shows burley in the overall production strategy of a farm. W.F. Axton's *Tobacco and Kentucky* (1975) provides a compact history of the state's tobacco production that is useful for understanding the relationship between burley and other types of tobaccos. There are also accounts of the production of Maryland tobacco that are somewhat similar to this book (Fertig 1986, McGrath and McGuire 1992).

The primary purpose of this book is to describe burley tobacco production in culturally grounded terms, but it also contains some explicit analysis expressing ideas associated with cultural ecology, a theoretical and conceptual viewpoint from anthropology.

This book is not intended to be a technical manual for the production of burley tobacco. To obtain the latest recommendations about effective practices, interested readers should contact their local representative of the Cooperative Extension Service.

1
Tobacco Culture

A burley tobacco culture exists. Men and women who raise burley tobacco share knowledge about tobacco production, a language with which they talk about tobacco, and even a sense of tobacco politics. Making a living by raising tobacco affects producers' "relationships to the land, their communities, their governments, and their churches, and how they are viewed as citizens" (Greene 1994, 71). While special knowledge and language develops around any agricultural commodity, some characteristics of tobacco have led to an especially elaborate "commodity culture." A discussion of these characteristics follows.

Tobacco is economically important in the places where it is grown. Like cattle in the old west, cotton in the Delta, or sheep in Australia, tobacco has a long history of centrality in the regional economies where it is produced. In these areas, tobacco is on people's minds because of its economic importance. The economic importance of tobacco in central Kentucky was mentioned again and again by those we talked with for this study.

"It's the main source of income," said Jimmy Bridges of Montgomery County. "Tobacco is your cash crop. You try to make ends meet with your livestock and grain, and if you break even and have a little tobacco money left over you figure you've done very well." Neva Greene of Montgomery County said, "[Tobacco] is really what most farmers around here rely on for their cash. The only thing I know of that would compare with it locally would be dairying, and that seems to be safe and it has been a pretty sure but very difficult way to make a living off a farm. Without tobacco, the average farmer here couldn't make it." Eugene Kiser emphasized the economic importance of tobacco in Bourbon County. "Tobacco was the mainstay for this particular part of the country; it was the mortgage lifter." In Oscar Richards's opinion, "It's bread and butter. It's the industry of the community, the state as a whole. It's the biggest industry in the state, the tobacco industry. You take tobacco out of Kentucky and this farmland wouldn't be worth a nickel. That's all that makes the farmland worth anything." Lucian Robinson summarized it this way, "Four years ago, I had a patch of tobacco over here. I had two and a half acres, possibly three acres. I came home from the warehouse [and] after my selling expenses and hauling was paid, I had $9,411 off of three acres of burley tobacco. That's a return that you can't duplicate. You can't replace that kind of income."

There is no substitute for tobacco in the economies of tobacco-raising families and their communities. For every million pounds of tobacco sold by the farmers of a community, more than a million and a half dollars are available for agricultural supplies, mortgage payments, food, new cars, and college tuition for the children. None of the widely grown row crops or forages comes near the economic returns of tobacco (Benson, Isaacs, and Trimble 1993). By the early 1990s, an acre of tobacco would produce gross income of $4,000, while corn produced $250 and soybeans $200 (Snell 1993, 73).

In public discussions about the future of tobacco, various alternative high-value crops are often mentioned. Experiences with these crops are mixed. Nelson Witt concluded that these alternative crops cannot produce the return of tobacco. "They can't raise anything on an acre of ground that will come up with tobacco. They've tried cucumbers, red peppers, pumpkins, cushaws, watermelons, and strawberries. They can't come up with anything that will make a third as much as tobacco. I had a little over fifteen thousand pounds this time. It brought $1.77. The check was about $24,000, but that's not all clear." A similar view was expressed by Willard Varner. "We tried them peppers back in '80. They tried to tell you that bell peppers would take the place of tobacco. That's impossible. But you get an acre of tobacco back there and get 4,000 pounds on it at $1.50, that's $6,000 an acre. You can't replace it. There is nothing to replace that. They talked you into raising these gardens, you have to raise a tremendous lot of all kinds of vegetables. That means hard work. You got to irrigate. You have to keep those vegetables looking good. With the least little fault, people won't buy them. You have to do it good peaches, apples, beans, and everything. Any kind of specks on them, people won't buy them. Tobacco is altogether different."

The economic importance of burley in this region has been declining as nonagricultural employment has increased, farmers have aged, and the future of burley tobacco has become more uncertain. In addition, burley production has gotten more concentrated in the hands of larger producers.

Tobacco has a long history. The crop has great historic depth in defined regions. Of all the economically significant commodities, it is the one with the clearest linkages to aboriginal times (Funkhouser and Webb 1928). The tobacco production technology in use today was originally derived from that used by native people (Greene 1994, 74). Tobacco is American. Some farmers see it as a link they have with Native Americans. In some regions of America, tobacco is a historic icon. The seal of the city of Lexington, Kentucky, has a tobacco leaf on it, for example. The tobacco leaf as a decorative motif appears frequently in our nation's capitol building in Washington (Berry 1993, 63). Long historical development is consistent with the elaboration of distinctive cultural systems.

Tobacco production is concentrated regionally. While burley tobacco is produced in fourteen states (Johnson 1984, 2), most production is in the burley belt, consisting of Kentucky and seven adjacent states (Haystead and Fite 1955). Kentucky is the largest producer of burley tobacco, and burley is the major cash crop of Kentucky farms (Census of Agriculture 1992). Burley tobacco became more important after World War II. Prior to this, farms were much more diversified, more self-sufficient and subsistence-oriented, and, therefore, much less dependent on tobacco.

Ira Massie, who farmed with his sharecropping father before a career in agricultural broadcasting and Extension at the University of Kentucky, described the position of tobacco. "It fell into place. You had your corn to do. You had your sheep to take care of. You had your cows to milk and hay to put up. So tobacco got its turn just like everything else. It didn't get the predominant care that you're giving it today. Tobacco gets first [priority] today."

Clearly, burley is important to the economy of Kentucky. Regional concentration of the production of a commodity fosters the development of cultural distinctiveness.

Tobacco is a fragile commodity and is difficult to process. Production of tobacco requires a lot of human contact and care. Farmers develop many ideas about tobacco because they spend so much time with it. There are many steps to the process, which means there is a lot to know. Each step may have an impact on quality; therefore, the knowledge required for doing each step right is important. This all forces a cultural elaboration. The relationship farmers have with small grain production, which is highly mechanized and less complex, is quite different.

Farmers have an open and noncompetitive relationship with each other in the marketplace. Unlike those of other specialized horticultural crops, tobacco prices are less sensitive to seasonal timing. We all know that the earliest tomatoes get the best price. Farmers producing for the potentially lucrative fresh fruit and vegetable market are in competition with each other. These producers can be quite secretive and noncooperative toward each other compared to tobacco farmers. The absence of this kind of competition among tobacco producers allows more sharing of information about improving production practices. Part of this lack of competition is due to the Tobacco Program's stabilizing effect on prices. The "tobacco program," a government-regulated marketing system, is discussed in Chapter 4. Tobacco farmers do not have to feel like their neighbors will outcompete them since they are all guaranteed a price. In effect, farmers who carry tobacco culture around in their heads have few reasons to not talk to each other about tobacco and will talk about it given the slightest encouragement.

Production practices have a substantial impact on quality. Tobacco producers have a strong commitment to producing a quality product. This is carried over to the marketplace in the form of the impact of quality on price. In most years, price is sensitive to product quality.

This theme is explored by the American poet, farmer, and tobacco grower Wendell Berry. "Burley tobacco, as I first knew it, was produced with an intensity of care and a refinement of skill that far exceeded that given to any food crop that I know about. It was a handmade crop; between plant bed and warehouse, every plant, every leaf, was looked at, touched, appraised, lifted, and carried many times. The experience of growing up in a community in which virtually everybody was passionately interested in the quality of a local product was, I now see, a rare privilege (Berry 1992, 85)." Berry seems to be saying there was a kind of ethos of quality in tobacco-raising communities.

Tobacco is marketed in a relatively processed state. There is a lot to tobacco farming. Farmers cure, classify, and package the tobacco prior to aging and manufacturing. Even after it is raised, there is a great deal to do. The different kinds of leaf produced on the stalk must be classified, for example. The produced tobacco leaves the farm in orderly, culturally significant categories. It is not milk or wheat commingled in a tank or bin. Among the last things the producer does in the production process is to attach meanings to the tobacco. Classification is a quintessentially cultural activity.

The botanical properties of tobacco shape the culture that produces it. Tobacco has an enormous leaf area, while its flower and seed are very small. The small seed size requires special seeding techniques, nursery beds, and transplantation. When growing, the leaf is a "livid green to bluish green" that changes to a lighter shade of green when reaching maturity (Garner 1946, 7). If the leaf is dried slowly after it is harvested, it turns yellow and then tan or brown. The fragile nature of the tobacco leaf and the way it slowly cures resulted in the construction of special facilities for curing and the knowledge required to manage the curing process on the farm. The flowers of tobacco vary from light to dark pink. While tobacco produces axillary buds (i.e., suckers), these remain dormant until flowers develop unless the apical bud is removed. This required the development of practices associated with removing suckers or preventing their growth with chemicals. Dried tobacco leaf can reabsorb moisture and become flexible again. This property relates to a complex set of terms and practices associated with maintaining the optimum moisture level of cured tobacco. When freshly dried, the leaf produces a harsh and irritating smoke if burned. When aged properly, this harshness is lost. This property is related to the post-harvest processing practices of manufacturers. It is important to its use for smoking that it burns with an even glow and gives off a fragrant aroma. One can say that if you grow a complex plant, you end up with a complex commodity culture.

A well-known botanical property of tobacco that structures tobacco culture is that tobacco contains nicotine. Nicotine produces pleasurable effects on the person using the tobacco product, be it smoked, dipped, or chewed. These pleasurable effects are subjectively described by tobacco users as satisfaction. It does not seem likely that persons would smoke if it were not for nicotine in spite of discussions about "taste" and "aroma." These subtle pleasurable effects are an important part of the value tobacco users place on tobacco. While it appears fair to say that tobacco users are after nicotine when they smoke, dip, or chew, they are also concerned about taste, aroma, and how a tobacco product represents an expression of personal style. That is, a transdermal nicotine patch and a Macanudo cigar are not the same.

"Nicotine makes tobacco addictive (Glantz et al. 1996, 58)." The use of the term addiction in the context of tobacco use is contested. Tobacco companies did not acknowledge tobacco as an addicting substance in public discourse until recent court cases, although purloined documents show manufacturer recognition of tobacco's addictive properties long ago (Glantz et al. 1996). Publicly, manufacturers speak of tobacco use in terms of a psychological process of habituation rather than a physiological process of addiction. Interestingly, the original surgeon general's report that spoke of the health risks associated with tobacco use spoke of habituation rather than addiction (1964, 3). Contested science aside, most smokers feel uncomfortable when they attempt to quit smoking after they have been doing it for some time. They report symptoms of withdrawal, such as irritability, problems in concentration, and anxiety. While it is probably true that some persons get less addicted than others, addiction is a widespread response to tobacco use. The addictive properties of nicotine are an important element in programs to make it more difficult and less attractive for children to smoke.

Addiction results in a continuing, inelastic demand for tobacco products. The pleasurable effects and the closely related pattern of addiction drive the entire tobacco production process. It does not seem likely that a significant number of people would buy tobacco products if they were not psychoactive and addictive.

Smoking is an efficient way of dosing with nicotine. The use of smoking to get nicotine results in two problems. The first is the production of harmful substances that are by-products of burning, benzopyrene and carbon monoxide among many other things. The second is the production of so-called environmental tobacco smoke (ETS).

Exposure to these substances through smoking and other forms of tobacco use is associated with a greater than expected occurrence of a number of serious health problems. These include lung cancer; cancer of the larynx, mouth, esophagus and other sites; chronic bronchitis; emphysema; hypertensive heart diseases; and arteriosclerosis, among other diseases (Eysenck 1986, 19). There are also associations, however weak, between disease and presence of environmental tobacco smoke (Redhead and Rowberg 1995).

The primary tool for examining the health implications of smoking has been epidemiological analysis. Epidemiologists make use of statistical analysis that looks at the ratios between the number of cases of a problem actually observed compared to what one would expect if there was no association between the behavior and the problem. This kind of analysis is concerned with "excess deaths" or "excess cases" and the "mortality ratio" (i.e., the ratio of observed to expected deaths) (Eysenck 1986). Kenneth Olden, director of the National Institute of Environmental Health Sciences, characterized this situation in the following way: "Much of the evidence we have that smoking causes cancer is circumstantial, although extremely strong (Bishop and Gyelin 1996, B1)."

While the actual mechanisms by which the use of tobacco products causes disease are not as well understood as the statistical association between smoking and disease, some recent research "provides a direct etiological link between a defined chemical carcinogen and human cancer (Dennissenko et al. 1996, 480)." The chemical carcinogen in question is benzopyrene, which is produced by burning tobacco and inhaled when people smoke. This research shows that chemicals produced from the metabolization of benzopyrene appear to bind with certain sites on the DNA molecule. These same sites are those "mutational hot spots" associated with cancer-associated mutations in a specific gene. When people smoke, they ingest chemicals including benzopyrene. When in the person's body, benzopyrene breaks down into another chemical called BPDE. Medical researchers found that BPDE tends to bind with the DNA of human cells. This did not occur randomly. Researchers found that binding was patterned, in that the BPDE was frequently attached to a location on a gene that suppresses tumor production associated with lung cancer-related mutations. This is the P53 tumor suppressor gene that made the cover of *Newsweek* Christmas week, 1996 (Begley 1996). Researchers concluded that the chemical metabolized from benzopyrene binds with and damages a specific gene's capacity to protect a person from lung cancer. This research was thought to provide "more definitive [laboratory] evidence" of how smoking actually causes lung cancer (Olden quoted in Bishop and Geyelin 1996, B1).

Public discourse concerning these issues is contested and disputed. The contested nature of the discourse can be seen in the proliferation of what appear to be semantic arguments. That is, health advocates, for example, take association as conclusive evidence of cause, whereas tobacco advocates often argue that cause has not been demonstrated. The statement that "association is not cause" is sometimes taken as pro-tobacco advocacy even when intended as a statement of scientific principle. Alternative models of cause are rarely taken seriously by health advocates. Information flow is impeded. This is especially clear with tobacco companies that have limited the dissemination of their own research when it supports the view that tobacco is addictive and that smoking is harmful. The contested nature of the discussion can be seen in the limited public discussion of the beneficial aspects of smoking in regard to self-medication to manage the impact of stress. Nevertheless, public, nontechnical assessment is consistent with the epidemiological view that tobacco use in general, and most specifically cigarette smoking, causes earlier death and increased disease.

There is, therefore, a sequence of causally related effects. Tobacco contains nicotine, which produces pleasurable effects and addiction. This results in a motivation to use tobacco, but more specifically, to smoke in a persistent way. Smoking means burning, which produces benzopyrene (and other by-products) which causes mutations in a gene that protects the body from the uncontrolled growth of tumors. Of course, the tobacco culture has been developing for hundreds of years in societies that do not or did not have a clear understanding of the health implications of tobacco use.[1] Increased understanding of the health consequences of tobacco use has had a significant impact on tobacco culture. The meaning of tobacco has changed.

While the controversies now associated with tobacco can be discussed endlessly, the focus of this book is the history (mostly twentieth-century history) associated with farming technology and marketing. To set the scene for our discussion, we offer this sketch of the emergence of burley tobacco, starting with the almost legendary account of its New World origin. We take the story up to World War I and the development of the modern American-blend cigarette. Elsewhere we discuss other parts of the burley tobacco story.

Burley tobacco is a variety of *Nicotiana tabacum*. *Nicotiana tabacum* is closely bound to the culture history of the United States of America. It was first domesticated in the New World and was in widespread use among Native Americans when Europeans first landed (Wagner 1971, 7). Tobacco became an important factor in the colonization of what was to become the United States as demand quickly increased after introduction to the European market in 1560 (Jahn 1954). By the beginning of the seventeenth century, consumer demand for tobacco in England was soaring in spite of the fact that most of the production was under the dominion of the Spanish. In 1610, John Rolfe emigrated to the struggling settlement of Jamestown, Virginia, and soon began to cultivate tobacco from *Nicotiana tabacum* seed obtained from Trinidad. Jamestown colonists were familiar with tobacco in England and were interested in raising it for their own use and as a marketable commodity (Bridenbaugh 1980, 35). As might be expected, the production system used in colonial Virginia was a development from Native American agricultural practices (Wagner 1971, 14). The cash crop became the basis of the economies of the thriving colonies of Virginia, Maryland, and the Carolinas (Main 1987, Jacobstein 1907). The flourishing tobacco trade attracted a steady stream of new migrants, mostly indentured males (Main 1987, 14). According to Main, "Tobacco is the reason [the immigrants] had come—as tobacco had paid the way for at least a third of all English immigrants to the New World in the seventeenth century" (1987, 9). Others argue that the unfavorable terms of trade and heavy debt burden that colonial tobacco planters had with English merchants and tobacco consignees were important factors in establishing colonial rebelliousness toward Britain (Breen 1985). It is clear that tobacco was an important factor in shaping the American experience.

    Tobacco was grown from the beginning of Kentucky's history to supply domestic needs and commerce. The production technology used in Kentucky was based on the system used in Virginia (Axton 1975, 30). Early evidence for tobacco production in Kentucky is found at Boonesborough. Historic records indicate that there were attempts to establish a tobacco warehouse as early as 1788. The Virginia Assembly authorized the establishment of a tobacco warehouse and inspection system in Kentucky and the appointment of official leaf inspectors (Axton 1975, 38) who determined if the tobacco was of sufficient quality to market. For this service, they received a payment of ten shillings per hogshead. Leaf that was inferior had to be burned. Upon inspection, the tobacco was warehoused and receipted. The receipts could pass for money at the rate of twenty shillings per hundredweight of tobacco. This system was developed to help assure distant buyers that the tobacco purchased was of acceptable quality. An archaeological survey of Boonesborough shows evidence of two tobacco warehouses used for the inspection required by the government and packing tobacco into hogsheads (O'Malley 1989, 21).

The development of tobacco commerce in Kentucky was inhibited by high freight costs going back east and trade restrictions downstream in New Orleans. Only light, valuable commodities, such as whiskey, ginseng, and furs, could be economically shipped up river (Baldwin 1941, 24). The trade restrictions imposed by the Spanish governor at New Orleans grew out of conflicts that existed between Spain and Great Britain. Tobacco did not begin to be a significant commercial crop in Kentucky until 1787 with the lifting of the Spanish embargo of New Orleans (Clark 1960, 87). The development of the New Orleans market for agricultural produce like tobacco, grain, and hemp was essential for Kentucky's emerging economy because it allowed shipment to Europe and the Atlantic coast (Clark 1960, 86).

By the middle of the ninteenth century, tobacco production was very important in central Kentucky. The kind of tobacco grown in this period was different from the kind grown today. White burley tobacco, as it is known today, evolved from a dark, air-cured type called red burley, originally grown by early settlers in Kentucky and southern Ohio (Gage 1933, Jones 1983). By most accounts, white burley was first discovered on the farm of Fred Kautz in Brown County, Ohio. Kautz's tenants, George Webb and Joseph Fore, ran out of seed during spring planting time. Fore crossed the Ohio and obtained seed from the farm of George Barkley of Bracken County, Kentucky. The plants produced with this seed exhibited unusual characteristics but were not transplanted and were destroyed. Saving some seed, George Webb raised it on his own farm the next year. Webb planted enough of the seed to raise twenty thousand pounds of the new variety the next year. An item in a Frankfort, Kentucky, newspaper by L.J. Bradford of Augusta, Kentucky, supplements this widely known, almost legendary account. Bradford claims that white burley was a product of a selection program he carried out on his farm "as early as 1860" (1873). He states that he gave seed to fellow Bracken Countian Barkley in 1864, and one of his own tenants planted six acres of the tobacco and sold the crop for an apparently good price. "It is a beautiful plant with a delicate yellowish green leaf, and white stem and fibers, cream color stalk, deep green," Bradford wrote. "When ripe and well handled, it presents the finest appearance of any tobacco in the world—bright as sunshine, transparent, clean, and clear of fuzz" (1873, 3).

The new kind of tobacco sold well, won prizes at fairs, and quickly spread through Ohio, Kentucky, and Indiana. It offered a number of advantages. First, being lower in natural sugars, it could more easily absorb flavors and sweeteners in processing. This offered considerable advantage in chewing tobacco, pipe tobacco, and, later, cigarette tobacco. The "dry" quality of burley meant that it was less susceptible to rotting and mildew. Second, it could be brought to market much sooner than the varieties it replaced because the entire plant could be cut at once. With the earlier red burley varieties, the crop had to be primed; that is, each leaf had to be picked as the plant ripened from the bottom to the top just like contemporary flue-cured tobacco. Curing itself was simplified as air-curing in specially constructed barns became the norm. Earlier varieties had to be heated to reduce their moisture content quickly enough to avoid rotting. In these ways, the mutation that produced white burley allowed changes in the production technology.

White burley is one of many types of tobacco raised commercially in the United States. These different kinds of tobacco are classified according to a comprehensive system established in 1929 by the U.S. Department of Agriculture. This system includes seven general classes: flue-cured; fire-cured; air-cured, with subclasses of light and dark; cigar-filler; cigar binder; cigar wrapper;

Tobacco Culture 13

and miscellaneous. These classes are divided into types and given a number and a name. The different tobacco types are suitable to different agronomic conditions and have different uses. In the U.S. Department of Agriculture classification system, burley tobacco is called Type 31, air-dried, light burley tobacco. This type is grown because of suitable soils and historical tradition. Roy Greene, a Montgomery County farmer with whom we spoke, said, "This area is adapted to this type, and burley was first grown around [central Kentucky]. Our season is ideal for that tobacco."

The demand for white burley tobacco increased dramatically with the development of the modern blended cigarette, first introduced by R.J. Reynolds Tobacco Company under the Camel brand in 1913 (Tilley 1985, 224). The Camel recipe made considerable use of burley. This, coupled with a well-financed advertising campaign, established Camels as a national brand and "paved the way for increased use of burley (Tilley 1985, 224)." The boom in burley for cigarettes was associated with a decrease in the use of it in chewing tobacco products and a change in the relative desirability of various types of leaf. Here we end the history of burley narrative for the time being. The historic thread is picked up at various places in this book.

2
Tobacco Ground

The process of raising a crop of tobacco usually starts with the decisions about where to place the crop and how much to plant. Because leaf is demanding on soil nutrients and vulnerable to plant diseases, these decisions are especially complex. The farmer must take into account soil fertility, topography, and the exposure to plant disease associated with a particular plot, as well as his or her own production goals and the relative importance of tobacco in the particular mix of crops for that farm. Decision making about placement of the tobacco crop often reflects a multi-year cropping cycle strategy. The range of possible cropping cycles available to the farmer has changed substantially through the years in which the people we interviewed have been farming. The key factor is a radical transformation in the farmer's capacity to maintain soil fertility through application of fertilizer. Chemical fertilizer use was infrequent before World War I, became more common between the wars, and increased dramatically after World War II.

Before the extensive use of chemical fertilizer, the best crops of tobacco were raised on land that was either newly cleared or planted after it had been allowed to revert to nature for a number of seasons since being planted to tobacco or other crops. That is, tobacco producers either used a true fallowing system, in which tobacco fields were not cropped for a number of years, or a multi-year rotation with as few as one crop of tobacco in it. This shifting cropping pattern is similar to that found in colonial Virginia in the seventeenth and eighteenth centuries (Isaac 1982, Jones 1956, Tatham 1800, Earle 1975). Simply put, without chemical fertilizer, the tobacco ground had to be given a chance to rest.

Fertilization allowed farmers to reduce the length of crop rotation cycles and fallowing. Finally, permanent cultivation of tobacco on the same plot became a technical possibility. Fertilizer also allowed farmers to go back to lands that had been long since depleted. In long-fallowed fields, then, there was a periodic process of clearing second-growth brush. This was called "grubbing out a field." In addition, some farmers reported that they cleared "new ground" for the first time. This meant that the trees and shrubs to be cut down were very large, requiring saws and axes. This was usually winter work.

Grubbing out a field was hard work. Arthur D. Jones recalls his experiences of clearing new tobacco ground when he was a young man in the mid '30s in Bath County. "Back then, a lot of ground wasn't cleared. My brother and father would take a grubbing hoe and axe for cleaning up the ground where they was going to put [a] crop of tobacco. Some ground was already cleared, but some they had to clear by hand."

The debris of the trees and shrubs from the grubbing process was bunched up and burned in place. There was no effort to spread the ashes although they would be plowed through. Or, according to Jones, "They take what they got off of it, and, if it wasn't too large, they used it to burn tobacco beds. If it was very large, it was used for [fire] wood." Stumps, not removed until later, would usually sprout for a few years, causing the clearing process to last for some time. Finally stumps would die and be taken away. The logs of larger trees might be dragged down into ravines on sloped lands. Some very old farmers recalled seeing large-diameter logs in ravines, indicating the debris of land clearing in the past.

Newly cleared fields had to be worked in special ways. There were specially designed plows for working a newly cleared field called "jumping cutter plows." These worked better than regular plows if there were roots to contend with. In the early stages of using a cleared field, the tobacco had to be planted with a hand setter. Production from "new land" brought a better price in the market.

In those days, some farmers used cattle in the clearing process. Farmers would, according to Paul Carraco of Carrollton, "pile up the brush around the edge of the tobacco patch and put it around there and get it up where cattle wouldn't come on through it. And then you'd have to put you a wire gap or gate between a couple of trees some place at the edge of it or set some posts there so you'd have a way to turn cattle in and out when you wanted them to graze on it. You built a fence around each little patch you had."

Even without fertilization, river bottoms or "overflow ground" could be cropped more intensively because of the periodic recharging of nutrients from flood-borne silt. Carraco described how the fertility of bottoms would be increased with sediment from the river floods. "The bottom land would flood [nearly] every year and leave maybe an inch, maybe two inches, maybe three inches, up to four inches—according to how bad the rain was that caused the flood and all—of sediment on top of the land. I still have this farm where I was raised, and this land has built up maybe three feet in some places in the lower spots where water had been over it and left the [sediment] in there and the dirt that washed off the hill land."

Many expressed preference for bottom land for tobacco if at all possible. In a multi-year cycle, this is not always possible. Ira Massie, well known for his tobacco broadcast programming, recalled his father's strategy. "We'd alternate the fields. We only had, I'd guess, seven or eight fields that we could use, and these were all the best fields on these farms [we rented on]. They were usually the bottom fields. Occasionally we'd have to go up on a hill because [of] our rotation system; we wouldn't have enough land to do [our rotation]. Generally we would rotate and stay in the very best land we had, then the bottom land at that."

Some farmers mentioned that the preferred placement of tobacco was on hillsides. This seems to run against common sense. It is interesting to try to piece together an explanation. Bottom lands represent a risk situation because of flooding. Tobacco is very sensitive to problems caused by poor drainage that could be avoided on sloping ground. Also, some felt that hillsides had the advantage of having more limestone in the soil. Limestone lowers soil acidity and makes more plant nutrients available to the growing plant. This land-use pattern was described by Paul Carraco. "Tobacco at that time was raised in hills, mostly. They'd get out on their hill ground, and you'd have what they called new land. Generally there was a good many rocks on it, and this rock produced limestone, and I think from this thing we got some value there out of our limestone where people in the bottom lands didn't."

The same variety of tobacco, raised on bottoms and hillsides, will produce quite different tobacco. A farmer active in Nicholas County, William A. Wilson, noted this in discussing his field placement preferences before World War II. "To have quality tobacco, they thought they had to raise it on a hillside . . . because the companies wouldn't buy it unless it was just the right quality." After reflecting that the companies did not pay much for even the best quality, he added, "They wouldn't hardly pay anything for redder, heavier grades."

In earlier days, the cropping pattern varied from locality to locality in terms of the length of the cycle, constituent crops, and specific cultural practices. Nevertheless, there is a basic pattern of tobacco rotating with other crops in a relatively long, multi-year cycle. Perspective on this pattern and the changes in it can be gained from the narrations of central Kentucky farmers.

Some farmers said tobacco always followed grass in the rotation. This strategy was used by Eugene Kiser, a Bourbon County farmer. "Usually you'd [have] five or six years before you'd plow a piece of ground. You usually plowed sod about every year for tobacco. And then that particular piece of ground would go into wheat and [then] go into grass. It would be six or seven years before you'd be back to it."

A tobacco, small grain, and grass rotation was widely used. Arthur D. Jones described this pattern. "If they had enough [land], it would be a while before they would get back to the same field. . . . I'd say at least four, five years, six more before they ever got it back again in tobacco. It's been that way ever since I have been cropping. Mostly they do rotate because if you're raising good tobacco, you mighty near have to do it. It seems like you might get by two or three years, but finally it runs it down. If you raise corn, you rotate it, just like you do tobacco.

You use your grain [wheat or rye] and you put your grass seed behind it."

The size of the farm had an influence on the structure of the rotation. Earl Jackson of Bourbon County described his early practices. "We usually raised [tobacco] two years in one place. After the second year we changed to a new field. Followed it up with a cover crop [such as grass, wheat, clover, or vetch]. [The rotation] depended on how big a place you had. If you had a small place, you plan on [having] about two or three places where you grow your crop and you would come back to the third field every six years. If you got a big farm you might put in a field no more at all [or] eight, ten years before you put it in again. You cut the wheat and stuff off of it and sell the wheat, bale up the straw, sell it for racehorse people for bedding. If you have a small farm, you'd have to rotate to go back to the first place that you grew at in six years."

Crop rotation cycles became shorter as soil fertility became more directly manageable by farmers. While these days the preferred production practice calls for the location of the tobacco crop to be rotated every year, several farmers we spoke with replanted the same fields for several years, replenishing some soil nutrients yearly with heavy fertilizer applications.

The process of intensification is described by Earl Jackson. "They do it a little bit different now. They usually have the land tested, and if you have too small a place, you can continue to use the same field over and over if you want to put the right kind of fertilizer on it. You can continue to grow it in the same place but you have to have your land tested to find out what the land calls for."

In any case it is recommended that tobacco be rotated with other crops for a variety of reasons. Earl Jackson described his approach to rotation. "[Normally] to start out with I always sow it in wheat in the fall, and in the spring I sow it in clover. Later in the fall I cut the wheat off it and bale it up for hay. And maybe later in the fall if the clover grows up enough I cut it off or if I don't, I cut it next year and bale it. And I sow grass with it, and it [will] take two years to grow up and [I will] let the stock graze it after that. I usually use fescue grass.... It produces pretty good. It makes a whole lot of grass. You can use the fescue, orchard grass, bluegrass, and clover. Of course if you mix grasses it covers the ground better, you get a better stand."

The nature of the total farm operation influences how the rotation is sequenced. Arthur Little discussed his experiences cropping on a horse farm. "At present I'm raising on three different farms. One of those landlords, he didn't rotate too often. He had a horse farm. A lot of these horse farms, they don't like to plow up the other ground. I don't believe that they rotate as much as the other people."

The best farming practice is rotation. Bud Rankin of Bourbon County advocated rotation in his account of contemporary rotational practices. "[Rotation] varies from one farmer to another. I have seen farmers raise tobacco continuously on the same field and still produce good tobacco, but your better farmers will rotate it every two years. Preferably, the ideal [is to] go to a new piece of ground every year. It's according to the size of your farm, but some people keep it out [of tobacco] five, six years. Some of them rotate it every other year. For a while I had one farm, I had three acres of tobacco and I had a six-acre field. I plowed the whole six acres every year, and I put half of it in soybeans and half of it tobacco. I would never take the soybeans off. I would just turn them under for fertilizer. Where I turned that crop of soybeans under, I put tobacco there, and next year I was growing soybeans [on the other plot] over here. You can do that on a small operation, and it will pay off. It wasn't a real fertile field and I was trying to build up the soil."

The tobacco patch is often put on the most fertile land with the best drainage. Roy Greene of Montgomery County said, "You usually select the most fertile land, and if you've been farming the same land over a number of years, you'll know where those spots are. And, of course, you consider the topography also. You want the land with as little slope as possible so you won't have erosion."

When deciding how much land to plant to tobacco, the farmer considers the amount he or she is authorized to sell through the tobacco program marketing quota. (The Agricultural Stabilization and Conservation Service (ASCS) office in each county keeps records of each farm's production and notifies the farmer each year of his or her quota.) The farmer must either attempt to grow that amount or plan how to alter it.

To improve the potential of the tobacco ground, farmers use fertilizer. Nitrogen, potassium, and phosphorus are the three major ingredients in fertilizers today. Fertilizers come in different analyses, such as 5-10-15 (read 5 percent nitrogen, 10 percent phosphorus, and 15 percent potassium). These elements may also be added individually if the soil is depleted of one ingredient in particular. It is possible to purchase fertilizer that is a custom blend of the three basic nutrients. Because nitrogen is an especially important nutrient in tobacco cultivation, it is sometimes supplied as ammonium nitrate. This compound, made from ammonia and nitric acid combined in water, has been used as fertilizer since 1926 and has the advantage of providing nitrogen at a lower cost than the balanced fertilizers. It does not contain either phosphorus or potassium.[1]

In the past, manure from livestock was saved and put on the fields. Paul Carraco noted that manure use has changed. "We don't have as much [manure] as we used to. With the advent of the big bales of hay, we don't feed as much in the barns. So you don't [obtain] as much manure to put on the ground, which is a great thing for the ground." As early as the seventeenth century, technical advice recommended against the use of animal manure because of the effect it had on the flavor of the tobacco (Main 1982, 31). This effect was demonstrated experimentally in the 1950s and caused the University of Kentucky to recommend limits on the use of animal manure on tobacco.

Before World War II, farmers who used chemical fertilizer applied very small amounts compared to current practice. Referring to his parents' Estill County farm, Nelson Witt said, "My mother would go along next to the row of tobacco, and she'd dig a hole out in line with the row. And I had a wooden ladle and fertilizer in a big bucket. I'd drop it in this hole. My sister came by with another hoe and covered it up. It came in an old cloth bag, like a burlap sack. That wasn't three elements like nitrogen, phosphorus, and potassium or potash. Most of that was phosphate. They used that and lime. That's the first they used on tobacco. They used that on corn before they used it on tobacco."

Others we interviewed mentioned the use of very light applications of fertilizers. Willard Varner stated, "I started using fertilizer back in the '30s. . . . We put a spoonful on each plant. It grew pretty good. Last few years that I raised tobacco, I put a ton an acre. It can be twenty-seven hundred pounds an acre by the time you put on the nitrogen and lime."

The economic benefit of fertilizer was clearly expressed in the advertising copy of the period. In an ad, a fertilizer dealer stated, "Here I think is a conservative estimate of the value of fertilizer on tobacco. We will take, for example, one acre of land that will produce 1,000 pounds of tobacco without fertilizer. This tobacco will probably net $.20 per pound, making $200 for the farmer from one acre. By using 150 to 200 pounds 3-10-1 fertilizer at the cost of $5.00 the same acre could be made to yield 1,200 pounds of tobacco and probably raise the quality to $.25 per pound, making the farmer $300 from the acre less the $5.00 paid for fertilizer. Think it over. I have plenty for sale" (Robertson County *Times Democrat*, July 5, 1924). Another newspaper note made the point that one farmer raised four successive crops on his land, and, with fertilizer, the fourth was the best (February 14, 1924).

By the 1930s the application rates had apparently climbed to three hundred to four hundred pounds per acre, still much less than the present usage. In the early days of fertilizer use, it was broadcast by hand. Later farmers drilled it in and cultivated the furrow next to the plant to get it closer to the roots.

As fertilizer application became more important farmers mechanized application. Shirley Wegner of Robertson County reported his early experiences with fertilizer application. "[We used] a one-horse spreader plow [to] furrow [the field]. We had a one-horse drill, and we'd put fertilizer in [it] and drill it right down in that row. You wanted a big shovel on your plow so you could get that fertilizer down pretty deep in the ground. It would spread right in the bottom of that furrow, and then [the drill] had two plows on the side of it that come along and ridged [the soil] up. You had a ridge to set tobacco on. Used it many a time, a one horse drill. We'd have to take one horse and lay it off, then go over the same thing with a drill, drill your fertilizer. You'd put [the rows] about three and one-half feet [apart] so you could plow them with horses." In a similar way, Arthur D. Jones described his first use of chemical fertilizer in the early '40s in Bath County. "We took and plowed this ground by hand. We hired a team of horses and laid this ground out in rows. Then they had a [single] fertilizer drill. You took a horse in this row and drilled the fertilizer. The first I can remember we used 300 to 350 pounds to an acre.

During the 1940s fertilizers were made available through the local committee that administered New Deal programs (i.e., Agricultural Adjustment Act Committee). Initially fertilizer was sold by stores in town. Now it is more typical for the fertilizer to be purchased from regional supply points and be delivered to the farm.

Bud Rankin described how they applied fertilizer in the early '50s in Bourbon County. "Back when I was a kid, we never broadcast fertilizer. We had a little thing behind a horse [that] put the fertilizer right over by the plant. It's what they called side-dressing, or putting it in the row. We'd run fertilizer in the row before we set the tobacco, and we would run it probably two times while it was growing right beside the row. [The machine] had a little wheel that metered the fertilizer out. It had two handles where the man walked and held them."

The relationship between soil acidity (pH) and plant response to fertilizer was understood early. Plants grown in soil derived from a limestone base responded better to fertilizer. Lime may be added to the soil to make nutrients more available to the plants by reducing soil acidity. Farmers reported that custom limestone crushing services were available in the 1930s. Limestone would be piled on the field and a rock crusher would be brought out to crush and spread it for a fee such as two dollars a ton. A note in the Robertson County *Times Democrat* mentioned that "nothing equals ground limestone as a fertilizer and many farms in the county badly need fertilizer. Tobacco is fast sapping the life out of our lands" (October 18, 1928).

Lime applications used to have some hidden costs. Kelly C. Haley of Bourbon County spoke of these. "Some farmers would have crushers, and they would just find rock around and they'd take these crushers and come to your farm and crush it for you. They did one terrible thing. An awful lot of our fences were crushed up. They didn't do too much blasting to get the rock out of the ground. A lot of good county cemeteries were ruined, like next door to us over there [it] had a beautiful rock fence. I guess it was several generations back that a man went in there and destroyed that cemetery. Then you had to take and put this crushed lime on a wagon and just throw it out with scoop shovels. They didn't have lime spreaders like they have now." It should be noted that through time Bluegrass farmers tended to consolidate smaller fields, thus making old rock fences surplus in an agricultural sense (Murray-Wooley and Raitz 1992, 132). Rock fences are a significant cultural asset in central Kentucky.

Now lime is hauled in on a bulk spreader-equipped truck and applied. A good time to lime the soil is right after the crop is taken off, to allow the lime to break down. Lime is available in various granule sizes. Fine-powdered lime is more expensive and has a quicker effect than coarser "ag-lime." High rates of fertilizer application today have increased the need to lime in order to neutralize the acidifying effects of the fertilizer.

Soil tests may now be done to determine the soil nutrients needed for the crop. Via local extension agents, soil is analyzed at the University of Kentucky, and the correct type and amount of fertilizer is recommended. Some fertilizer companies also do soil testing. The main purpose of soil testing is to reduce fertilizer costs, as Allen J. Whalen of Bourbon County said. "When I started making the decisions, I started using soil tests a little more frequently to find out what I really ought to be putting on, and trying to save a little money as I go along."

Most farmers stated that soil should be tested every year to determine fertilizer applications. However, after recognizing the efficacy of this recommendation, some indicated they did not have their soil tested each year. Oscar Richards stated, "You are supposed to have your soil tested, but I never do. I pretty well know my ground because I tend it so much, you know, and pretty well know what it needs. If it needs more nitrogen, I'll get a high amount of nitrogen in the fertilizer, and if it don't need too much of that, I'll get a higher analysis of potash. But it is better to have your ground tested, and it will save you some fertilizer for sometimes you miss it. You misjudge it and [that] causes extra fertilizer to be bought."

Roy Greene pointed out that excess applications of phosphorus are not really wasted as it remains in the ground for the following years' crops. However, he felt that if one uses too much nitrogen, the plants never mature. They keep trying to grow with a heavy, green crop resulting. Ideally, the plants run out of nitrogen just before the end of the growing season and are mature for harvest.

Fertilizer is ordered for delivery in early May for incorporation into the fields. This may be purchased bagged for application by the farmer or in bulk. Bulk applications are usually applied by the dealer. Some farmers are suspicious about the analysis of the fertilizer and the weight. Anecdotes about fifty-pound bags being two pounds short are common. Fertilizer has been under some sort of regulation for a long time. Legislation passed in 1886 required that fertilizer manufacturers have the University of Kentucky Experiment Station analyze the content of fertilizer sold in the state (Smith 1981, 28-30).

With chemical fertilizers, higher yields are possible. Previous to heavy fertilizer use, Roy Greene raised twelve hundred pounds per acre. Now his yield is twenty-seven hundred to three thousand pounds per acre. Willard Varner indicated that his yields were about five hundred to six hundred pounds an acre before fertilizer and that they could go as high as four thousand pounds with fertilizer. While heavy fertilization produces high poundage, some farmers feel it can hurt the quality of the crop.

3
Tobacco Labor

Labor is one of the main expenses in tobacco production. Farmers have always had four ways to obtain labor for their crop. They and other family members can do the work; they can share or swap labor with neighbors; they can hire people; or they can have someone sharecrop their land. In the past, more labor was provided by members of the family and cooperative work-swapping groups formed of neighbors (Rosenberg and Coughenour 1990, 1). The loss of these sources of labor is keenly felt during the unmechanized, labor-intensive steps in the production of the burley tobacco crop. The reduction of the availability of labor from family and neighbors relates to out-migration of rural residents, increase in the average age of farmers associated with different household composition, and changing relationships with neighbors, among other factors.

Under current circumstances, many farmers must hire people during the periods of peak labor demand associated with setting, housing, and stripping of tobacco. There has been little historic change in the pattern of peak labor demand in tobacco production. The tobacco production cycle has three periods of high labor demand interspersed with periods in which work is less demanding and more attention can be directed toward the farmer's other crops or his off-farm employment. The first high-demand period is transplanting in May. While there is plenty of work to do in June and July, it is something a person with a tractor, cultivator, and sprayer can manage. Beginning in July with topping, labor demands increase. Cutting and housing tobacco during August and September is a second period of high labor demand. While stripping tobacco in November and December is less pressured than spring planting and late summer harvesting, it involves a great deal of work.

Other components of the farmer's operation are part of the tobacco labor equation. George Duncan, a University of Kentucky agricultural engineer, commented on how some of the different commodity combinations work with tobacco. "Tobacco and dairying has not been a real compatible combination (even though a lot of dairymen and a lot of tobacco farmers are one and the same) because the dairy is a seven-day-a-week, twenty-four-hour-a-day job nearly, so a lot of them do not want to have to take away from that management of the herd to produce tobacco. Beef cattle is a good combination [with tobacco] because you can give and take, and the beef cattle can go ahead and take care of themselves generally for three or four days a week while you're working in tobacco."

The most labor-intensive aspect of tobacco production is the harvest, which starts in August and continues through September. In central Kentucky, the weather is hot and humid during this time. Most harvest labor is hand labor. Milton Shuffett, an agricultural economist who specializes in tobacco, described the situation. "[There has] always been a problem with harvest labor in tobacco. It's hard work, awfully hard work. It all has to come in a short period of time. Once that tobacco matures and the weather is right, the farmer is very anxious to get it cut and housed in a very short period of time."

There has been little progress in mechanization of this part of the process. A farmer's perspective on the harvest and its mechanization is provided by Alex S. Miller. "There's always a labor shortage about [the] time to house tobacco. That's the big expense. People at the university keep working on ways to mechanize it and cut down on this hand labor. But really there [hasn't] been a whole lot of progress made. They've come up with machines that'll cut it or tried to—I've never seen one work—and they put it on wagons, different wagons, different way[s] to move it to the barn hoping they'll save some labor. But it still takes a whole lot of labor to raise an acre of tobacco." The tobacco harvesting machines

Tobacco Labor

Miller referred to are expensive. Informants said the machines could cut approximately one acre per day, which is comparable to what can be done by hand. "It's not that much faster, but it's a lot easier. It's extremely hard work to cut tobacco," said Jimmy Bridges.

In the past, a much larger portion of tobacco labor was provided by family members and work sharing between neighbors. Evelyn Toadvine, recalled that little was spent on labor when she was a young woman in Mason and Bourbon counties, "The whole family went out. You didn't hire too much labor when she was a young woman in Mason and Bourbon counties. Once in a while they'd have maybe one or two men come from the mountains and they'd stay with us to house tobacco and strip tobacco. Once in a while you'd trade labor with a neighbor. We had a good neighbor lived right across the road from us and we'd always trade."

An implication of the situation described by Evelyn Toadvine is that women were involved in aspects of tobacco labor. Indeed, women have always been involved in tobacco production. Geneva Witt of Bourbon County did tobacco production work as a child and as a wife and mother. She described an incident from her childhood after World War I. "I [worked in the fields] when my daddy was living. He asked [my sister] and I to get out there to chop the weeds out of it. He said, 'Be careful or you'll cut it down.' So I cut down a big plant. Stupid like, I stuck it back in the ground, I thought it would grow and he wouldn't know the difference. Soon as he went out there, that was the first thing he found. So I can remember chopping weeds out of tobacco pretty well." She recounted how she has done many different tobacco production tasks on the farm that she and her husband Nelson bought and developed. "I've helped here with farming ever since we have been farming [starting in 1945]. I've always helped with every bit of it. When he went to the field, I went to the field. Lot of times I'd go pull plants. . . . And maybe there would be three or four here for dinner, and I'd fix a batch of dinner for all of them. And soon as they's over with, I'd go back and pull plants. And when we got enough pulled then we'd set. I drove the setter while they set." Her daughters also helped with field work. "My daughter, the one in Florida, Betty Jean, helped with the plowing. When I wasn't there, she was. Yvonne, the youngest girl, worked less on the farm, but she'd take the mowing machine and go mow hay. She'd mow it off clean and good. When she was little, I'd take her to the stripping room and put her in a box with a pillow. That's where she slept. So when the boys come along, I took Ricky to the stripping room and I took Malcolm. That was the second boy. And he took pneumonia and so that was the end of taking the little ones to the stripping room. I didn't do any more after that year. After they got in school, I went back and I would help strip and do whatever had to be done."

The wives of tenant farmers are more likely to be directly involved in tobacco production than the wives of landowners. Many tenant farm wives work full-time in local factories and businesses to supplement the farm income and therefore cannot do much farm work. But often the wives and children of tenants do work during peak labor seasons, especially housing and stripping. They may also help with the mowing and driving the tractor during setting. Neva Greene described this situation and how it is changing. "[Tenant farmers], the ones who are living on the land and raising the tobacco for half, their wives and children used to help more with the tobacco crop. But now so many of the wives work in the stores and work in the factory, you will hardly find a woman at home on the farm during the day. I think it is sad, but I think it is true. I was talking to [a neighbor] the other day and she said she didn't think it was necessary but desirable if you want the things people in town have had. There has been a change in the last twenty or thirty years."

Women had a major responsibility in feeding hired hands. Allen J. Whalen made this clear when he said, "They had five and six men that worked full time and stayed upstairs and worked every day and fed them three meals a day and did their laundry. And all that sort of thing. It was a lot of hard work then for the women. In doing tobacco work, we would often be served dinner. In most cases, it was a very well-prepared meal of meat, numerous fresh vegetables, and biscuit.[1] In the few cases where we did work for farmers whose wives worked, we ate a simple lunch of bologna on light bread."

Labor swapping was also important in earlier times. The nature of labor swapping is described by Roy Brown of Brodhead, Kentucky. "We did a lot of swapping work. Farmers would swap out and help one another. The farmer's tobacco may be ready to be cut, and his neighbor would pitch in and help cut his, and then get his housed. They'd come over and then cut the other neighbor's tobacco when it would get ripe enough to cut. It took several hands to hang tobacco, and if you could swap work, it just saved hiring somebody and it reduced your costs of raising a crop."

Work swapping had its own etiquette. "In those days if you were working, your neighbors come to help you," explained Eugene M. Kiser. It was common knowledge that you were going to eat dinner wherever you were working. If the dinner wasn't ready, [Dad] would help Mother put the dinner on the table. In those days, everybody just worked together. If [someone] had something to do, if he could use you, you went over there. Time wasn't taken care of. You worked by days then. There wasn't such a thing as an hour any more. I don't know whether there was real tight reins kept on how much you helped so-in-so or how much time they helped you. You just got the job done; that was it."

Swapping could be extended to equipment according to Nelson Witt. "I've run a high-boy for thirty years. We don't do commercial spraying [with it]; we swap with neighbors. If I need a haybine and don't have it and this fella has one and let me have it. [We] swap back and forth with the machinery. That way we don't have to license it and go into extra cost."

As mentioned earlier, the extent to which family and neighbors can be relied on as labor sources has decreased. As family-supplied labor and work-swapping with neighbors has decreased, hiring of labor has increased even though labor supplies have become tighter. Milton Shuffet, an agricultural economist, described the current situation. "I am sure it's more of a problem now because there are not as many rural people as there were. Families are not as big as they were, and there aren't many kids around to help in tobacco. [There is a] lack of workers that are willing to sweat and work in tobacco. The wages it takes to pay those that are willing to still do it compared to working at Toyota or the filling station or McDonald's or some of the other places that young people and other people can work part-time [are high]. The factories that have come in to provide employment for folks that did not want to do the farm work. And the minimum wage that has pushed some of the wages up, certainly for people to have a decent income, but it has caused the farmer to lose some of his labor supply. It has put him in a much tighter situation on being able to afford some of the labor costs. So the labor is what is pushing the farmer to look at new ideas like never before."

Farmers say finding good labor is a problem. Oscar Richards discussed this. "The people's got too ornery to work, and farming is hard work, and you can't go out and hire a hand to work in tobacco hardly at all. Now since we've had such big unemployment, it is a little easier to get help, but it is awful hard to get good ones." People who haven't worked as farm hands seem to discount the skills and commitment necessary to do many of the tasks for which extra hands are hired. Oscar Richards emphasized the need for people with farm backgrounds. "Now you have heard that farmers are dumb. You can go out here and take a farmer, a man off the farm, say thirty years old that has been a farmer all his life, you can take him to a tobacco field, he'll make a top hand just like that for he is ambitious [and] eager to work. You can go to town and take somebody that's raised in town and been there 'til age thirty, you can bring him out here on the farm, I'll guarantee you he'll never make a top farm hand. There is no way he'll ever make one. It's just too hard to teach him. There's things that he'll never learn. They're simple to me. But he's not going to put that much in it to learn."

There were significant changes in the composition of the rural population after World War II. The availability of locally resident, rural farm labor decreased as people migrated to take up industrial employment. Silas Cleaver of Bourbon County described the transformations he has observed in agricultural labor since the early 1950s. "[The laborers] lived off the farm. Most of them had houses in Millersburg, which is about two and half mile from here. They would walk there, then walking soon became out of style and we would go pick them up in a pick-up truck, and they usually brought their lunch and stayed all day. And then times got more prosperous, and they drove to work. And then our labor situation seemed to change. Most of the Black people seemed to acquire different type of work. They left the farm and moved to places like Middletown, Cincinnati, and Columbus, Ohio, and started working in industry. Then [the] labor [situation] changed, and we mostly used all white people. And that went along, and we would have pretty steady the same people, some of them living off the farm. They were dependable and worked good on the farm up 'til the time of, I guess you would say, when Lyndon Johnson was president and we started having the Great Society, food stamps, and more welfare. Slowly farm labor has diminished to hardly any. Right now we [are] using for our major amount of farm labor Mexican migrant workers with the exception that I have one fellow that's worked for me for the last eleven years. [He] has been very dependable. We can no longer find the labor here because they can actually make a better living, as they call it, doing no work on welfare and food stamps than they can working on the farms. So they will no longer hardly work, or if they do come out to work, they're not dependable and will work a half a day and then come back in two or three days and work again. You can't depend on them as a source of harvesting a crop or putting a crop in the ground."

The theme that "welfare" somehow spoiled hired tobacco hands is frequently expressed by older farmers. One can assume that "welfare" payments of various kinds and policies about earned income and eligibility would change a potential employee's decision making about a particular job. Another factor, of course, is the increase in opportunities for nonfarm employment in factories within and near Kentucky's rural counties.

Allen J. Whalen, reflecting on farming in the early 1930s, noted that labor problems have resulted in diminished care in producing tobacco. "People don't spend as much time on tobacco now as they did in 1950, '51. They don't take as much pains with it, I don't think. That's partly because help is not as plentiful now and the cost of labor has gone considerably higher than it was back then. Everbody's cutting corners on everything that they try to do to maximize income and try to cut the overhead. The price of the labor and housing the tobacco is nothing like what it was in 1950. We had two or three full-time employees in 1950 at five dollars a day and we'd furnish them a house. You're lucky if you can get that at five dollars an hour now and furnish the house."

Because of these developments, migrant labor has become much more important. Milton Shuffet, commenting on alternatives to mechanization, indicated the increasing significance of migrant labor. "The alternative to mechanization has been migrant labor. The southeast tobacco states, flue-cured tobacco, began using migrant labor ten or fifteen years ago, because they first came into the Florida vegetable production and, of course, migrants have been a great mainstay of the labor force in California and Texas. And even Michigan and Indiana have used migrants in vegetables and orchard harvest for many years. But Kentucky has been resistant to that type of labor force. [A few] years ago, it began to happen, and now it's becoming more prevalent for farmers to arrange to get migrant laborers to help with the harvest, housing, and sometimes the stripping of tobacco in the fall. So that has been an alternative to mechanization because it doesn't require a big long-term investment. It's a headache that comes and goes each year for three or four months, and if you get through this year then you can start over next year and make another decision of what you want to do about migrants and labor. And in a lot of cases they have been of equal cost or less cost than local workers and are often times twice as productive. Farmers explain how they can get so much more work out of migrants [than local workers] just because the migrants are here to work, to make [as] many dollars as they can [to] send back home to the families or [take] back when the season is over."

Since 1988 the number of migrant farm workers involved in tobacco production has increased substantially because of limitations in the supply and reliability of labor (Rosenberg and Coughenour 1990). Estimates for the number of migrant farm workers in the state was two hundred in 1989 (Rosenberg and Coughenour 1989), seven thousand in 1994 (Rosenberg 1994) and ten thousand in 1997 (Carlton and Estep 1997). These workers are part of the seasonal migrant labor stream that flows south from Florida and north to sites in Michigan and Wisconsin. Most of these workers are Hispanic, and many are legally certified by the Immigration and Naturalization Service under the provisions of the Immigration Reform and Control Act of 1986 to work in the United States. The current involvement of Hispanic migrant farm workers started with vegetable production and extended to tobacco.[2] Cutting, housing, and topping tobacco are the dominant labor activities done by these workers on Kentucky farms. The top burley-producing counties have the most Hispanic migrant farm workers (Rosenberg 1994). Mark Reese, agricultural Extension agent for Scott County, said "We couldn't get the crop in the barn without migrant labor" (Carlton and Estep 1997, A11).

Farmers who hire these workers need to adjust to their limited English language skills and need for housing during their stay. Some farmers develop stable long-term relationships with certain farm workers who return from year to year. Some workers are able to stay all year and make use of local community education and social service programs. Public sector programs have been supplemented by both Protestant and Catholic church ministries and secular groups, such as the Kentucky Migrant Network Coalition. In the past few years, there has been an increase in businesses focused on the Hispanic market. Most visible are the grocery stores and restaurants found in many central Kentucky cities that address the food preferences of the emerging Hispanic community.

Landowners often rent the use of their farm for a share of the crop. Owners may own more land than they can farm. Sometimes a woman inherits a farm from a husband or parent and cannot raise the crop herself. Older farmers may also choose not to raise their crop. In such cases, the landowner may make an arrangement with persons willing to grow the tobacco on shares. For fortunate landowners, the tenants are often men who have worked on the farm for many years, together with the owner. They can place complete trust in their honesty and judgment.

If the tobacco is raised on shares with a tenant, or farm manager, the economic arrangements can be complicated. The tenant-landlord relationship is one of mutual dependence. Arrangements vary from farm to farm, but generally the landlord provides the land, marketing quota, and facilities such as barns, a stripping room, and tobacco sticks. The landlord purchases the fertilizer for the tobacco crop and may provide some of the machinery as well. The tenant is responsible for providing labor (either his own, his relatives', or hired hands'), machinery, and chemicals as needed for the crop. Most important, the tenant and the landowner split the gross sales fifty-fifty.

Born in 1930, Evelyn Toadvine recollected her father's experiences. He raised tobacco "on the shares" in Mason and Bourbon counties in the late 1930s and early 1940s. "He didn't have to pay any rent, he just raised the crop for them for half of it. . . . We didn't have to pay rent. The house was furnished. That was the deal back then. Well, still is in a lot of places. They furnish you a house to live in and you raise the tobacco for the half of it. He just mowed the fields and whatever had to be done, bale hay and things like that for himself and the landowner, too. Only thing he got was half the tobacco, and he had to pay all the expenses. The landowner would furnish fertilizer or something for bugs."

Berle Clay described the situation of the tenants on his father's large Bourbon County farm. "You have to recognize that each of those families was given a house and a lot. And he probably had a barn, and he could also raise five acres of corn, and he could probably have, it would be specified, up to eight or ten [shoats] and a milk cow and so on. So they had a large subsistence operation of their own too."

Sharecropping families made good use of the farm to produce food for their table and some production for sales. Arthur D. Jones made this clear as he reflected on his experiences. "We'd kill all our meat there. I killed about four hogs. We had a garden and all the milk we needed. Back then, in the '40s and early '50s, you could sell anything mostly. We even sold cream. He built me a hen house down there. Had a little chicken house, raised young chickens in it. Had a whole bunch of eggs. [We would] sell eggs [to] buy groceries with. We didn't buy no meat at all. Once in a great while you might buy a little bologna. 'Cause you had your ham and your shoulder. And you'd have the sausage hanging up so long. You'd hang them up till warm weather starts. You couldn't keep them. You have to eat them up. They canned sausage [and] other parts of the meat. And cured the side meat too. All the chickens you want. You could kill hen or roosters in the winter time. Boiled them. Very little you bought. You had sugar you had to buy, and flour, salt."

When asked to identify the characteristics of a good landlord, Shirley Wegner of Robertson County said he would be just like his landlord of the past twenty-five years, "I done just as I wanted to with the farm after he bought it. He'd sometimes wouldn't know where I was gonna plow. He'd ask me where I was gonna plow. And they'd been people trying to rent it from him, but he'd say, 'Wegner ain't gonna leave me.' I've lived with him twenty-five or thirty years, I don't know just how long it was, but he was just the same one day as the other. We never had a word of falling out, no way, shape, or form. Never a cross word. He'd pay for fertilizer to put on my tobacco, and when I tended corn he'd furnish everything, tractor, gas, and all the equipment and fertilizer. And I'd get a fourth of the corn then. Give him the rest. I'd do just whatever he'd want me to do. I never turn him down when he asked me to do something for him. That's the way we done each other. He'd help me. He'd come to me and we'd be housing tobacco or something like that, busy, and he'd be helping us and he'd say, 'reckon you can get along without me tomorrow?' . . . He'd want to go somewhere. And I'd say, 'Yes sir, go right on.' But he'd come and ask me, and him being the landlord. He was a real fellow. Now, there wasn't nobody better than Ches Reed. There might have been people just as good, but there was nobody better. I always felt that I had enough experience that my decision was as good as anybody's and I'd like to use my own judgment. With Reed, everything [was] left up to me. Whatever I wanted to do, he thought it would be all right. That's the way we done. Whenever I did anything that Ches would get mad at, he'd never show it. Always my decision, he'd look for me to make the decision. Of course he never had no whole lot of experience about farming no way. And he really didn't know how to farm. And he left it all up to me. I've seen landlords that you couldn't please no matter what you done. I raised [him] a good crop of tobacco and I paid for this farm two or three times in tobacco for him. He only give $7,500 for this farm when he bought it. And I paid for it in no time. And he told people that I was the best tenant that he'd ever had or ever would have."

Wegner contrasted this landlord with others. "I had one landlord that I didn't get along with. Well, in fact I had two that I didn't get along with. I did for a while, but the old fellow that I lived on over here, well, I had a crop of tobacco out, come a big washing rain and washed it about all off. It did wash off places as wide as this house, that there wouldn't hardly been a plant left. Well, I got in there and reset that tobacco, and plowed it, and he didn't want me to plow it. And I told him he had his living made, and I said, 'I've got to try and make my living here.' And I said, 'I'm going to plow this tobacco.' And he told me that if I was plowing it, then I could hunt me a place another year. I said, 'Well, I can do that.' I said, 'I was looking for a place when I come to get this one.' Well, I told him that, 'If you'll get out of this field where I'm raising this tobacco, go back to Mt. Olivet and let me alone, we'll get along. But I'm going to try to raise some tobacco.' And I did. I plowed it out, reset it, and had a pretty good crop that year. And when time come to rent, why, he came down. I knowed what he was up to. And he tried to get me to stay. I said, 'No, you told me I could hunt me a place, and that's what I done. I done rented this place here.' He wanted me to come and get off this fellow and stay with him. I said, 'No, I don't do thataway.'"

In Alex S. Miller's experience, tenants took good care of the crop. He said, "We were blessed with good tobacco men. You didn't have to worry too much about how they took care of it. You knew. When I was running the farm, that was one of the least worrisome jobs I had. I knew that the people that were raising the tobacco were going to take care of it. Frankly, they knew more about raising tobacco than I did, and you didn't have to worry about but what they'd give that tobacco crop of theirs tender loving care. I knew they would."

Berle Clay reflected on his family's experiences in the 1950s with a large Bourbon County farm that required the help of a number of farmers working through a sharecropping arrangement. "At that point we were farming about sixteen hundred acres. The only other people working on the farm were sharecroppers. There was no sort of managerial level person. My dad did all of that. The people who weren't working in tobacco the rest of the year was doing things for the farm like planting corn, harvesting corn, working with the cattle, working with the sheep, and so on. Farm labor was an additional thing that the tobacco sharecroppers did. [All of the tobacco was sharecropped]."

The changes in the rural population that influenced availability of labor had a similar impact on the tenant farmer population. "The size of the labor force has just shrunk tremendously [since then]," Clay said. "When we moved to the farm, there was about six families living on the farm besides us. Back in my father's time, it would probably be double that. In his time, the labor force was largely, or if not entirely, Black. But I have the impression that [when] tobacco sharecropping developed in the first part of the 20th century, the bulk of the sharecroppers were white. So I think in the period from, say, 1910 to 1940 or '50, you got a gradual displacement, maybe rapid displacement, of Blacks leaving the farm. And certainly in the late '30s and during World War II, there was a general migration of Blacks off the farm to other areas in other cities, such that when I was a kid I can remember only two Black sharecroppers and I think they were out of sharecropping by 1950. And they would have been the last. With the exception of a man who milked the cows and mowed and did things like that, they would be the last Blacks living on the farm. So there's been a change in the composition of the labor force."

"When we moved there, the other work of the farm, the non-tobacco work, was done by the tobacco tenants. They had a certain time off to work on their crop. [We had temporary help] irregularly. At that point there wasn't that much pressure on getting the work done. I remember distinctly one family. We had one family in this period who did nothing but labor. We hired the father and three sons. And I remember them picking up hay bales with a team of mules and a wagon and carrying them to the barn."

4
The Tobacco Program

Burley tobacco has been sold through a federally supervised marketing system since 1933. Growers who participate in the Burley Tobacco Program are authorized to sell a certain number of pounds of tobacco at or above a minimum support price during each marketing year. The total number of pounds that can be sold in a given year is determined by the U.S. Department of Agriculture through a formula that accounts for domestic demand, export demand, and reserve stock levels (Snell 1996). Tobacco taken to market under this program is graded by specially trained U.S. government tobacco graders. Each of the 115 government grades that may be assigned is supported at a fixed price determined on the basis of the quality and demand for the grade. If the companies do not bid at least one cent more than the support price for a graded lot of tobacco, the tobacco "goes to the pool." The pool is managed by the Burley Tobacco Growers Cooperative Association located in Lexington, Kentucky. Price support programs in Kentucky, West Virginia, Ohio, Indiana, and Missouri are administered from this office. There is also a co-op in Tennessee for Tennessee, Virginia, and North Carolina. When tobacco goes to the pool, or co-op, the farmer receives a payment for it from the warehouse like it has been sold to a buyer. The warehouse is reimbursed by the co-op, which uses money lent from the Commodity Credit Corporation, a federal corporate agency that finances and otherwise manages a number of commodity programs. The loans plus interest are paid back when the tobacco is later sold to a manufacturer or tobacco dealer. Virtually all costs associated with this program, including interest on loans, storage costs, and grader salaries, are paid for by tobacco producers and purchasers. Some administrative costs are paid for by taxes. These include market analysis, research and Extension, crop insurance subsidies, and administration of the program by USDA (Snell 1996).

The Burley Tobacco Growers Cooperative Association does not actually take title to the tobacco as it remains the property of the farmer. The pool acts as an agent for its member farmers to store, grade, process, market, and ship the tobacco acquired through the tobacco program. They never physically possess the tobacco.

After the sale, pool tobacco is shipped from the sales warehouse to a processor nearby. The tobacco is redried and put in hogsheads or other containers and stored until sold. Tobacco manufacturers can place orders with the pool for various kinds of tobacco of various crop years. Recently the pool had tobacco six years old. Some crop years, virtually no tobacco goes to the pool.

Jimmy Bridges detailed this system. "The government keeps this floor on the tobacco, the pool price.[1] We know we're gonna get that for it, right around that. The government'll give you that for it, so if the buyers want it they have to give you just a little more. Sometimes they'll give you five, six, seven, eight, maybe ten cents more than the government puts on it."

The Burley Tobacco Program has effectively maintained adequate and stable prices for this agricultural commodity since its inception. Development of the contemporary tobacco marketing system started with the increased concentration of tobacco manufacturing interests in a few companies during the late nineteenth century. These included the British-owned Imperial Tobacco Company, the European state-owned monopolies, and the American Tobacco Company (Axton 1975, 87). The most dominant force in the tobacco market was the American Tobacco Company, formed by James B. Duke in 1889 (Campbell 1993, Durden 1975, Winkler 1942). The "trust," as it was called, was "a huge combine of the principal eastern tobacco manufacturing companies (Axton 1975, 82)." By 1910, the trust produced 86 percent of all cigarettes in America as well as large portions of plug, smoking tobacco, fine cut, snuff, and little cigars (Tennant 1971, 27). It also had large ownership interests in cigarette production machinery, container production, retailing, and tobacco flavoring compound firms, which fortified its strategy of increasing market dominance.

Through the trust, Duke worked to radically reduce the number of competing companies buying tobacco in order to control the market. Duke, because he was well capitalized, was able to gain market share through aggressive price cutting that reduced competitors' cash flow. He then bought or merged with the competition as their financial positions declined. This domination was aided by control, through exclusive contracts, of the newly developed, highly efficient Bonsack cigarette manufacturing machines (Tennant 1971, 41, Winkler 1942, 79) and elimination of independent tobacco buyers.

At its peak, American Tobacco had an ownership interest in 250 companies (Tennant 1971, 27). Without competition, farmers received lower returns on their crop, sometimes not enough to cover costs. Farmers are quick to recall stories from their fathers and grandfathers about selling tobacco for less than the cost of taking the product to market, and not being able to sell some grades of tobacco at any price. In addition, tobacco was purchased in the barn by agents of the trust rather than at public auction, and if a farmer did not consent to selling he might not be able to sell at all.

Paul Collins of Mayslick describes the marketing situation under the conditions of the trust. "We had special tobacco buyers that would come around. I remember when I was a youngster, maybe six, seven, maybe eight years old, a buyer, Mr. Tom Malone, lived down here in Maysville, came to the barn. Dad had asked him to come out and look at the crop. They came around once you got through stripping and looked at it after you had it in keeping order. So he looked around. 'Well,' he said, 'Tom,'—he knew my father—'I'll give you six cents for the tips and seven cents for the red and throw the flyings in.' That was before we had cigarettes; you just threw your flyings in." (Tips, red, and flyings are types of leaves, which are explained further in Chapter 9.) Tobacco sold in the barn was taken to a tobacco receiving point. Reynolds Bell mentioned, "I can remember [my father] taking me to a tobacco receiving point. They would put [the tobacco] in a hogshead, and then have a screw that pushes a header down on the tobacco in the [hogshead], presses it together so they can get 'X' number of pounds in it. At that time, that screw was powered by manpower, and men pulled walking around in a circle which was screwing this header down on the tobacco."

Prior to World War I tobacco farmers began to work to increase prices. Their efforts represent one of the few cases in American agriculture where farmers were able to improve prices through organizational efforts (Saloutos 1939, 1960). The story of these developments in the western Kentucky dark-fired production region are well documented. In 1904 some farmers in the Black Patch region, the western tobacco district of Kentucky and Tennessee, organized the Dark-fired Tobacco District Planters' Protective Association. These farmers attempted to pool their crop through a sales cooperative in order to control the supply and increase the market price. At the height of these efforts, 70 percent of western Kentucky tobacco farmers belonged to the association. The tobacco companies dealt with this by offering better prices to farmers who had not signed up with the association. Nonparticipating farmers were the targets of terroristic attacks by what came to be called Night Riders (Nall 1939, Kroll 1965, Miller 1936, Warren 1939). Ultimately Night Riders attacked the buyers, warehouses, factories, and rehandling facilities of the large tobacco companies. There were highly organized raids on facilities at Princeton, Hopkinsville, and Russellville (Axton 1975, 92-93).

Similar developments occurred among burley producers. Local and then regional farmer groups were formed in the Bluegrass. Central Kentucky saw its share of Night Riders (Campbell 1993, 98-116). While much less has been written about the "tobacco wars" in central Kentucky than about similar events in western Kentucky, the efforts in the Bluegrass appear more successful and somewhat less violent.

A masked Night Rider. (J. Winston Coleman Kentuckiana Collection, Transylvania University Library)

During the 1890s, demand for tobacco was increasing while prices were declining. This ironic development was attributed to the lack of competition in the marketplace brought about through the efforts of the trust. Initial organized activities included the creation of county-level farmer groups. In early 1898, a Carroll County, Kentucky, farmers' group encouraged state and federal officials to "enact more stringent antitrust laws against industrial combinations (Campbell 1993, 99)." Local groups also appeared in a few other Bluegrass counties. Burley producers from seven counties met in 1898 and sent petitions to the U.S. Congress and the Kentucky legislature. Though the leadership of this group began to advocate withholding of tobacco from the market, no action was taken. An organizer from Shelby County unsuccessfully attempted to create a manufacturing facility through the establishment of a joint-stock company to be owned by growers. By the end of the year, there had been a meeting of burley producers in Lexington with representatives from twenty-four counties. This organizational meeting resulted in the formation of the Kentucky League of Tobacco Growers.

This grower group was in favor of reduction of tobacco production by its members. It also attempted to stimulate legal action against the trust, recruit more farmers to the organization, and establish independent warehouses. The league failed quickly, but participants gained leadership experience that served as a foundation for later developments. This experience demonstrated the immense difficulty in raising capital to build independent, cooperatively owned tobacco warehouses as a mechanism to raise tobacco prices.

New efforts to organize producers began in 1902 with the creation of the Farmers and Tobacco Growers Association in Carroll County. Its program was based on the idea of advancing payment for tobacco to farmers and storing the crop until prices went up (Campbell 1993, 104). The Burley Tobacco Growers Association, established at Lexington during the same year, attempted to control the market through contracts with producers. The association was successful in contracting for a large portion of the crop and arranging for its sale. One of its leaders got as far as meeting in New York with James B. Duke, although he rejected the association's offer. This effort failed because of insufficient capital and the active opposition of the trust (Bleidt 1932). Similar attempts to pool the crop in 1904 and 1905 failed because of the inability to raise money to finance farmer payments through bank loans or issuance of stock.

The 1906 crop year brought some success. The American Society of Equity, established in Indiana by James A. Everitt to organize Midwestern wheat farmers, met with a group of farmers in Henry County. This resulted in the establishment of what was known initially as the Henry County Union. This group met with similar organizations from Grant, Owen, Pendleton, Shelby, Spencer, Trimble, and Washington counties, creating the Burley Tobacco Society (Campbell 1993, 109, Bleidt 1932, 25). Organizers who were part of similar earlier efforts in western Kentucky assisted the organizational efforts of the Burley Tobacco Society. Organizers worked to sign up farmers throughout much of the burley production area. Headquartered in Winchester, Kentucky, this group pooled the 1906 crop but was not able to arrange a sale to the trust.[2] This impinged on the group's ability to borrow additional money to finance the pool and created hardship for its members. In spite of its failure to sell the tobacco, the group succeeded in getting commitments to pool from farmers representing more than 50 percent of the projected 1907 crop by the first of the year. Based on this, the Burley Tobacco Society decided to attempt the pooling of the 1907-1908 crop under the leadership of a wealthy grower, Clarence Lebus. Larger landowning growers were among the first to sign up (Campbell 1993, 111-112). This participation pattern helped in the acquisition of loans from bankers. The banking community saw that increasing farm prices would increase commercial transactions in their markets and improve the value of their loan portfolios (Campbell 1993, 112). By summer, the society had signed up farmers representing about 75 percent of the crop. Again the society attempted to bargain with the trust. Unfortunately for the society, the trust had substantial amounts of burley in storage so the pooled crop was, like the 1906 crop, not sold. As a consequence, the society's financial assets were tied up in the stored tobacco of two crop years.

Out of the society's success in pooling and failure to sell came a proposal which escalated the conflict between society and trust. The leadership proposed that its members not plant any tobacco the next year. There was, in fact, an extensive boycott during which production fell by more than 95 percent (Campbell 1993, 126). There was a difference between the earlier pooling efforts and what came to be called the "cut-out of 1908." Although violence was denounced by the leadership of the Burley Tobacco Society, the cut-out was enforced by the Night Riders through beating noncooperating farmers, destroying plant beds, destroying non-pooled tobacco in warehouses, and sending intimidating anonymous letters to farmers (Cunningham 1983, Bleidt 1932, 34). Violence and intimidation were added as an "organizational incentive." The intimidation is illustrated by anonymous notices published in central Kentucky newspapers that stated, "Dare to raise a stalk of tobacco on your land or assist any one else in the raising and ye shall pay the penalty with your home and life. So be ye warned" (Campbell 1993, 123). While intimidation was part of the story, most violence was directed at crops not at people.

Paul Collins of Mayslick explained aspects of the work of the Burley Tobacco Society.[3] "[The society] was an organization for the poorer people, the people that were tenants and made their living principally by growing tobacco. The people who were pretty affluent didn't want to join in. They wanted to sell their tobacco wherever they wanted to. [The purpose of] the equity was to get [a decent] price for your tobacco so that you could make a living out of it. Back in 1907, I think, was when the equity started. There was several counties [represented], Fayette, Bourbon, Montgomery, Clark, Nicholas, [and] Harrison. They wanted to hold the tobacco until they got a living price for it. [The society was successful] for a while. I think the trouble was financing. They did not have enough financial heft, you know, to make the thing go. That was back in the time of the Night Riders. If a fellow didn't join the equity, they went out to see him, you know. Took some whips to him. Of course that wasn't right, but that was the way they tried to do it. I knew plenty people that was in that. I had a brother that was in it. They would visit these fellows that wouldn't come in and grew tobacco on the places and give them a little lashing."

Christine Sims of Robertson County remembered a relevant event from her early childhood (around 1908 to 1910). "We happened to live right close to the actual work of the Night Riders. We lived up on the top of a hill, and one of our close neighbors lived down in the flatland around us, and he didn't sign up in the pool, and the Night Riders visited him. That caused a lot of hard feelings that I can remember in the neighborhood. It was a local young man that raided the old man, took him out and whipped him and destroyed his tobacco bed. It is still remembered by their descendants and played a vital part in our church. [The Night Rider] led the singing and played a very important part in the church. If I am not mistaken, my father wouldn't go to church for a while after that. The [whipped man] was not real integrated into the community and I guess that's why they picked him out because all the rest of his neighbors joined the pool and he wouldn't, and he was going ahead with raising his crop and so the Night Riders visited him."

The cut-out worked. The trust purchased the crops of 1906, 1907, and 1908 at the prices bargained for by the Burley Tobacco Society. Dwindling supplies in the hands of the trust and the society's sales of tobacco to independent manufacturers were factors important in achieving this outcome (Campbell 1993, 128).

Even with the success of 1908 behind them, the program of the Burley Tobacco Society was not sustainable, and the crop of 1910 was not pooled. Although the society was "effectively dead after 1910" (Campbell 1993, 144), the trust paid higher prices that year. During these times, the trust was also taken to court. In 1907, the U.S. government started legal proceedings against it through the provisions of the Sherman Anti-trust Act. Finally, in 1911, the U.S. Supreme Court directed that the trust be broken up, much like Standard Oil had been subdivided earlier and AT&T much later. A number of new companies were created: Liggett and Myers, P. Lorillard, R.J. Reynolds, and a smaller American Tobacco Company. In its opinion, the Court stated that the American Tobacco Company was acting to drive its competitors out of business (Robert 1949, 152).

World War I had an important effect on burley tobacco. Paul Carraco reflected on the effects of the war and some of the postwar efforts to improve commodity prices in tobacco. "In [the] war years, 1917 and 1918, tobacco was being used more then than it'd ever been used before. The soldiers were using a lot of cigarettes, and they were chewing a lot of tobacco. And they seemed [to] like that. They [got a] little tension relief by smoking and chewing, and it also seemed like it was good business for these tobacco companies to send cigarettes to the army. So with the advent of the World War I and what it brought on, in 1917 tobacco really got high and got up to a dollar a pound in 1918."

The high prices did not encourage the development of farmer marketing organizations. As is usually the case, farmers responded to high prices by producing more. As Paul Carraco said, "And they were wanting tobacco real bad. Well, the season finally ended, and by the time that '18 crop was sold, whereas part of it had brought a dollar a pound, it was selling for eight or ten cents a pound. They had enough tobacco raised and our old supply and demand entered into again. They had enough tobacco to keep all the tobacco companies in tobacco and then some. It made it a rough game. Our people went ahead and continued to raise it, but when the supply and demand got like it was you didn't get enough out of tobacco. People would go out and put out twice as much tobacco as they had this year to pay off their debts for next year. First thing, you got so you couldn't give tobacco away. [It] got down to the point where [around] 1920 it'd be about eighty cents a hundred to sell it. Maybe you'd go to the market and you wouldn't have an offer of anything for it and so they'd just knock it out to the house. Sometimes people would come along later and [buy] up this tobacco and give practically nothing for it. People wanted to get some way they could sell their tobacco and keep this tobacco at a good price where people could afford to raise it and all."

Tobacco prices collapsed along with the prices of other agricultural commodities. Opening tobacco prices in the 1920 selling season were half what they were the previous year (Campbell 1993, 153). Overproduction did not seem to be the cause as the 1920 crop was actually considerably less than the previous year's production. Severe credit tightening by the Federal Reserve motivated by the goal of inflation reduction, was a major factor (Campbell 1993, 153). The collapse of tobacco prices brought with it farmer agitation in Lexington, causing the tobacco market to be closed for a time.

Farmers again turned to the alternative of pooling. "At the end of 1920, people began [to] talk about a pool," Carraco continued. "Jim Stone [of Lexington] and Ralph Barker of Carrollton made speeches all over the country and got the leaders out of the counties, and they got a pool. A pool was a pooling of tobacco, of throwing all tobacco together in their grades, and a certain price for each grade, and striving for this price. Maybe they had four different grades of lugs, bright leaf, red leaf, and tips. They were all priced out according to the price that they'd been bringing."

This pool came to be known as the Burley Tobacco Growers Cooperative Association. The impetus for this came from Robert W. Bingham, owner of the Louisville *Courier-Journal*. Persons involved in the association included Aaron Sapiro, a California lawyer with knowledge of cooperatives; J.C. Cantrill, a member of the Kentucky congressional delegation; Bernard Baruch, a New York financier; and leading growers and warehousemen such as Stone and Barker (Axton 1975, 103-104, Robert 1949, 202). The association attempted to sign burley producers to five-year contracts and provided for paying farmers for tobacco when it went

This photograph of a tobacco payment check for 10 cents was published in the Burley Tobacco Growers Cooperative Association's magazine, *The Burley Tobacco Grower*. It appeared under the headline "Check Tells Story of Another 1920 Crop Tragedy." It was sent to the association by a grower who stated that it represented a half share in payment for a one-thousand-pound load of tobacco. At the request of the grower, the editor published the photo for the lesson it taught about farmers not marketing cooperatively. Stories about such low payments were common among the people we interviewed. (University of Kentucky, College of Agriculture, Library)

to the pool based on the average price of that grade (Bleidt 1932, 54). Buyers then bought tobacco from the association, which then settled up at the end of the season. This required a large amount of capital. Although it appeared to be in better financial condition than earlier organizations, the association was chronically underfinanced. Some of the money borrowed to finance association operations could not be used for advances to farmers, and much of the money borrowed carried a very high interest rate. Legal threats to have these arrangements declared in violation of federal antitrust legislation reduced the association's access to much needed credit (Bleidt 1932, 53). In the face of this, Robert W. Bingham lent the association $1,000,000 (Ellis 1982).

In spite of these problems, the organization had accomplished a great deal by 1926. They claimed 109,000 members and had a grading system in place. State legislation had passed that

The Tobacco Program

required landlords to have their tenants deliver the crop after the landlord had signed a contract. In addition, the association had developed a program focused on health, recreation, homemaking, and education through a women's auxiliary (Bleidt 1932, 58). It also published a magazine for its members.

While in many ways the association was successful, there remained the problem of under-capitalization. From the farmers' perspective, payments for tobacco they delivered were often delayed. They may have drawn a portion of the proceeds for their tobacco but had to wait for much of their money until the crop had actually been sold. These installment payments were called "draws."

Paul Carraco experienced draw payments when he was a boy. "Well, I remember I was a kid, I think, eleven years old, and I had a patch of tobacco. It was a pretty tall tobacco, stand-up leaves, and a pretty broad leaf and we took it to town. They was going to have a draw, so much on each pound of tobacco that you raised. Well, I think I had about eight or nine hundred pounds, something like that, and we took it to town and put it through the market, and my first draw on it was fifteen cents a pound. So that was really good. And so they took this tobacco then and stored it in different warehouses. They'd receive and sell tobacco. You were supposed to have three draws, and if you got fifteen cents on the first draw and you'd get fifteen cents the next draw. And you might get ten cents [on] the last draw, and you might not get five. You might not get anything."

The tobacco manufacturing companies' attitude toward the association and its participating farmers has been described as one of "implacable hostility (Campbell 1993, 154)." Companies were reluctant to buy pooled tobacco. This, coupled with limited association capacity to pay farmers, meant farmers began to break their contracts with the association.

A farmer in considerable need of cash might easily be tempted to sell his tobacco outside the pool to a "dump house." Paul Carraco described these operations. "Dump houses were tobacco warehouses that sold tobacco all the time, and maybe the pool just had 80 percent of the people signed up, and this other

Published in *The Burley Tobacco Grower* in the 1920s, this political cartoon addresses the cooperative's problem with non-members who benefitted from the better prices brought about through the association's market organization work. Other cartoons depicted non-pooling free-riders as backward, old, and unprogressive. (University of Kentucky, College of Agriculture, Library)

Still Milking Our Cow Through The Fence

percent may have figured that they could go out and do what they wanted to. But when the companies got to getting in there and bidding in these dump houses over and above what they would get, what the people who were in the pool were getting maybe on the first draw."

In the face of this, the association released farmers in 1925. The crop of 1926 was not pooled, and prices fell. It was as if the association made the tobacco price problem worse, for in the late 1920s the prices of other agricultural commodities were relatively high. The association failed because it could not consistently pay advances to farmers, it could not prevent free-riding non-poolers from benefitting from its efforts, and it was not able to control production adequately. Historic interpretations of this episode vary. Axton's view is that it represented an important learning experience and more or less showed what was necessary for a program to work (1975, 105). Campbell expresses a much less optimistic view: "Experiments in cooperation over the past twenty-five years [preceding the Depression] provided no clear-cut alternatives" (1993, 154).

The Tobacco Program 51

Farmers continued to have significant problems marketing their tobacco. For example, farmers were often prey to speculators, called pinhookers, described here by Paul Collins. "After they started selling at the warehouses, these fellows had money and all and was pretty well-to-do. They'd go down there [to the warehouse] and meet those fellows at the door when they were bringing the tobacco in. [The producers] wouldn't know what the tobacco was worth, what it was bringing. [Pinhookers would say] 'You know, you have a pretty good crop here. I'll give you twelve dollars a hundred for it.' So naturally a lot of them would sell, not having to pay warehouse fees and not being sure what it would bring. The pinhookers done pretty well because they knew tobacco. Fellows wouldn't pinhook if they didn't know what it would do. A fellow that was a pinhooker was Ruben Toad that lived just on the edge of the county over toward Fleming. He went out to see a big, rough-looking fellow, who we called Dutch. He said, 'That's a pretty good crop you got there, Dutch. I'll give you so much for it.' Dutch said, 'Well you've bought it.' Darned if he didn't make as much money on that crop as Dutch did. It was fine tobacco. They'd come out and try to buy it at the barn. [This lasted until] they got the co-op organized. After they got the federal government ahold of it, you know, and made the price legal like we have today. When they organized this thing as it is today. That killed the pinhooker because the farmer was certain about what he was going to get. A lot of them had been connected with warehouses and knew about selling price. After this became organized by the federal government, you [could] grow only so much. Some farmers wouldn't pay attention, wouldn't listen to the [pinhookers] at all, especially a fellow that had been on the market any himself. He'd just drive by them. He knew better. But some poor fellow that hadn't been down on the market, maybe from another county way back up here [didn't know]."

It took the Great Depression to create the opportunity to solve the chronic price problem of tobacco. The current tobacco program grew out of New Deal legislation, although many of its features resonate with earlier community-based efforts. In response to the Depression, Congress passed the Agricultural Adjustment Act

[AAA] in May of 1933 as part of Franklin D. Roosevelt's eventful first hundred days of New Deal legislation. The goals of this act were "to establish and maintain such balance between production and consumption of agricultural commodities as will reestablish prices to farmers that will give agricultural commodities a purchasing power with respect to articles that farmers buy, equivalent to the purchasing power of agricultural commodities in the base period" (quoted from Daniel 1985). This legislation focused on wheat, cotton, field corn, hogs, rice, milk, and tobacco and provided for restricted production and benefit payments to the farmer.

The act established the concept of parity price as a key goal of U.S. government farm policy in the post-war period. Parity refers to the "ratio between agricultural and industrial prices" as these existed in 1919 (Irons 1982, 111). The concept is arcane but reflected the experience of American farm enterprises after World War I. There was a shift in the relationship between commodity prices and production input costs. Farmer "purchasing power" declined by 45 percent from 1919 to 1933. There was not just a shift in relative prices but a transformation of the countryside. Production technology improved. Inevitably, with this improvement came increased costs for machinery, improved seed, and other inputs. This investment raised production and, as a consequence, prices declined. Ironically, farmers suffered from their own increases in productivity.

Parity prices for tobacco were based on a later time period than other commodities. Tobacco-state politicians managed to establish the period of higher prices during World War I as the reference period rather than the prewar period used for other crops to the advantage of tobacco producers. Tobacco prices had slumped badly after the war and, consequently, using the same reference period as used for the other agricultural commodities would have been a hardship.

The Agricultural Adjustment Act provided for the reduction of production of the commodities covered by the law. Initially farmers were allowed to produce a defined number of pounds of tobacco. In 1940 there was a shift from a pounds-based system to

The Tobacco Program

one based on acreage (Tennant 1971, 190). By the 1970s it was changed back to pounds again.

The program was voluntary. Farmers who did not participate were free to sell their tobacco and would benefit from higher prices. Potentially, therefore, there was a free-rider problem. This was dealt with through a tax on the sales of nonparticipating farmers, which made raising tobacco outside the program economically unfeasible. The program did raise tobacco prices.

In addition to a concern for commodity supply and price, legislators turned to the markets themselves. The U.S. Congress enacted the Tobacco Inspection Act of 1935. This led to the development of official tobacco grade standards and a mechanism for tobacco inspection and grading and typing. In addition, there was legislation to reform warehouse practices, including licensing of weighmen, inspection of scales, increased time allowance for farmers to reject bids, and maximum speed of auctioning tobacco.

United States v. Butler, a test of the constitutionality of the Agricultural Adjustment Act, came to the U.S. Supreme Court in 1936. At the same time, other precedent-setting New Deal legislation was being tested legally. The Court reached the opinion that the act was unconstitutional on the basis of a number of issues. The act provided that the costs of raising commodity prices to parity would be paid for with money derived from a tax charged the first processor of the commodity in question. This important provision was rejected. The majority saw it as "expropriation of money from one group for the benefit of another," rather than as a tax collected to enhance the "general welfare." There was also concern that farmers were being coerced to participate in the programs of the AAA and that the government should not be involved in agriculture. The disposition of this case was an important component of the larger political struggle between New Deal proponents and the opposition. This is still being played out in the contemporary debate over farm policy and the goals of downsizing big government.

The production control functions of the legislation were carried forward by the Soil Conservation and Domestic Allotment Act of 1936, passed a month and a half after the constitutional demise of the AAA. The Soil Conservation and Domestic Allotment Act allowed production control by treating the overproduction problem as an environmental issue. That is, it defined tobacco as a soil-depleting crop, the acreage of which needed to be controlled (Daniel 1985, 128). In 1938, a constitutionally cleaner version of the Agricultural Adjustment Act was passed.

The economic benefits to farmers were expressed through minimum or floor prices for tobacco grade by grade mentioned earlier. The system of marketing quotas had to be ratified by vote of two-thirds of the growers. A farmer could sell more tobacco than his marketing quota, but this was subject to large penalties (Robert 1949, 211).

The tobacco program is crucially important to tobacco farmers throughout the burley production area. Montgomery County farmer Neva Greene summarized her view of the program. "Tobacco became a safe and secure crop after they started the tobacco program with an allotment for each farm and a government support price under it, which didn't mean that the government paid you for raising tobacco but only that they guaranteed a certain price for a certain grade. Then if the companies didn't buy it, the Burley Tobacco Growers Co-op took it—they called it 'under loan,' but we always assumed our tobacco was sold when they took it—and they re-dried it and stored it and sold it later. And if they made a profit after paying the government what they had borrowed to maintain the program, then they divided the surplus again with the farmers whose tobacco they had taken under loan.[4] So that was a support system that our economy needed."

Paul Carraco described this further. "Now, the pool that we have, the government is behind with a support price. Of course, that has worked down to the point now where we pay our own day-to-day expenses on it, and the government, it's not costing them anything, really, to administer the program. But we do have the fact that the government tells you how much you can raise, and that in turn has worked back with the use of tobacco in the United States and what it's been used for and how much they can expect to use next year, and how much they can expect the year after next, and how much they used three years ago, and can keep your supply and demand pretty good." Organizations involved in administering the program include the county Agricultural Stabilization and Conservation Service (ASCS)[5] office, the warehouses, the Burley Tobacco Growers Cooperative Association, and the companies. The production control program is administered on the county level by an Agricultural Stabilization and Conservation Committee. Although originally each county had its own ASCS office, there has been some consolidation. Each office has a committee elected through voting districts. Each district elects three members to the Farmers' Committee (membership of 15), which in turn selects three of its members to form the committee. The duties of the committee are to administer the government agricultural programs and to adjudicate problems. Each committee meets in the spring. Apparently, in earlier days, the county committees dealt with technical production recommendation matters on seed varieties and fertilizers as well as marketing program administration. Within limits, they can adjust marketing allotments. This helps to ease resentment within the community over unequal distribution of tobacco allotments. They also deal with grievances and problems, such as farmers who have oversold their quotas. Most important, they coordinate the allocation of next year's marketing quota as determined at a national level based on the past year's production.

Farmers have voted to renew the program every three years since it was initiated. As Oscar Richards said, "They vote three years at a time. It is something that people will never vote out. If they have an election on it, there won't be more than one or two in the whole community that vote against it."

There were many problems with this system at first. Sam Whaley of Robertson County reported a disagreement he had. His fields were measured in his absence, and the surveyors neglected to subtract the portion of the field under a large tree. This produced an "excess" of tobacco, and they cut it down before he had a chance to argue the case. Arthur Harney Jr. felt the acreage measurement system was not reliable and said, "[There] used to be people come out to the farms and measure our tobacco, and you'd get some people [that would] measure it liberally, and you'd get others that were real conservative. It was kind of like playing Russian roulette with who measures your tobacco."

The acreage-based system created an opportunity to produce more pounds while the market created incentives to do so. Roy Greene described these problems associated with a quota based on cultivated area rather than marketed weight. "We were on an acreage program. There was no limit to what some farmers were going to grow to an acre and they were overproducing. They had to keep cutting acres to keep production in line, so then they came up with the idea of saying how many pounds you can grow instead of how many acres. You can grow as many acres as you want to but you can sell only so many pounds. It varies from farm to farm. You've got a basic quota and they work from that. If there has been an excess of tobacco that has been grown in the past, they reduce the total poundage a little bit, and if there is a shortfall, they will raise it."

Techniques farmers used to increase yields included closer planting and heavier applications of fertilizer and other chemicals. Under this system, much tobacco had to be acquired by the pool and stored. Overproduction in the past had been dealt with by reducing the acreage allotments.

Acreage reduction became ineffective as a management tool because there was a provision in the acreage law that established a minimum allotment of a half acre. As more allotments came within the minimum, the production controls that were so important to the function of the policy did not work as many producers did not share in the reduction. Conditions were moving toward the point that all growers would have a half-acre plot. The most marginal producers were affected least. For example, in 1955 there was a referendum of burley-raising farmers to approve a reduction in market quota. Larger farms' allotments were to be reduced 25 percent. Farms with six-tenths or seven-tenths of an acre were reduced one-tenth of an acre, and farms with half an acre would remain the same.

This production control policy caused problems for the owners of larger farms. Berle Clay reflected on his father's experiences as owner of a large farm in Bourbon County. "They kept cutting back the number of acres and where that hurt was with the large farms. There was a lower end for the small farmer, I think it was under an acre, where they made no reductions in their acreage. But the cuts all came out of the large farms. So we always complained, you know, that [a] Tennessee hill farm producing less than an acre of tobacco was producing the bulk of the tobacco, whereas the acreage was being cut down on the large farm. As a matter of fact that's why my dad bought [a new] farm at Little Rock. [He did it] to get the additional tobacco acreage to add to the farm quota in view of these cuts in the acreage. Now it's on pounds. It's different."

At that point, tobacco in storage amounted to 2.8 times the annual use. Substantial losses were anticipated in price support operations. There was widespread feeling that the program could not survive. Granville "Gus" Stokes, a University of Kentucky College of Agriculture associate dean at the time, called interested parties together to try to solve the problem. With the help of Senator John Sherman Cooper, they got legislation passed in 1971 that put the poundage system into effect.

Granville Stokes discussed the development of this new policy. "We were overproducing. The acreage program provided a system of control that allowed the farmer to overproduce. The yield per acre was going [up.] The acreage was going down, and the production remained constant. Tobacco under the acreage program was continually in trouble with overproduction because it offered the farmer an opportunity to use better techniques and beat his neighbor. His big job was to have the highest yield in the neighborhood."

Overproduction led to dramatic increases in the amount of tobacco in the pool. Concerns for the future of the tobacco grew. Facilitated by the University of Kentucky, leaders in the tobacco trade began to meet to deal with the program's problems.

"In 1969, we went up to the Tobacco Workers' Conference, and every paper that was given up there was gloom and doom," Stokes continued. "I never have been to a scientific meeting that was more demoralizing than that particular meeting. There was nothing positive about it. We had 450 million plus pounds in the pool. How were we going to get out of the box? I met with Ernie Hillenmeyer, who's the president of Parker Tobacco now, talking about this terrible situation that had developed in the economic section of this meeting that we were in, not knowing where to go. And I said, 'Something has to be done.' And so when we got back [to Lexington, we] had hearings.

"I [organized the hearings] because I didn't know what to do and somebody had to do something, so I said, 'Well, this college has to do this.' And so I proceeded and went ahead with it and they all showed up. And I just said, 'Now, this is what I see,' and just laid it out how terrible a shape we were in, overproduction burden, need to do something. There was about ten or fifteen of us that kept meeting time after time after time. Louis Ison, president of the Kentucky Farm Bureau, gave us the breakthrough, I think, that caused that effort to turn into legislation that became the poundage program. And it was the elimination of the acreage requirement. And the legislation was passed. I think it either has a late December or an early January date of 1971, and the '71 crop went into the field as a poundage crop for the first one."

The amount of pounds a farmer can sell changes based on the supply situation from year to year. The original allotments were made on the basis of how much tobacco was being grown on the land in the 1930s. It is difficult to obtain marketing allotment for land not part of the original allotment. A tobacco marketing quota adds a great deal to the value of real estate (Gibson 1964, Shuffett 1986, 9).

The farm's marketing quota is tangibly represented by the marketing card. It is described by Philip Barry Sims, who served as the executive director of the Harrison County Agricultural Stabilization Conservation Service office. "It's [like] a regular credit card, just the same size. [It] has the farm name, the owner's name. Each farm has a serial number, and they're all on the marketing card." The number of pounds the farmer can sell off a specific farm is recorded on each card. Sims noted that each farmer "has to present it to the warehouse when he delivers tobacco for sale."

The warehouses are involved in the administration of the poundage system. The farmer's crop is weighed when he delivers it to the warehouse. The weight of the delivered tobacco is deducted from the farmer's marketing quota. If a farmer does not produce enough poundage to fill his quota he is "down in weight," or below his allotment for that season. He may try to make up for this by growing more tobacco the following year. He may "carry over" up to 100 percent of his quota (Atkinson et al. 1981). If a farmer overproduces, he is allowed to sell a certain percentage over his poundage allotment, provided that he subtracts that percentage of quota from his marketing card in the next season.

Virginia Calk explained this system. "We have marketing cards that tells you are allowed so many pounds each sale, and you can raise 100 percent of your allotment. And you are allowed 110 percent[6] in one year, but if you raise 110 percent, that 10 percent is deducted from your allotment the next year. Or if you are under your allotment, they add that onto another year, makes it (a) better base. Some years we come up short. This past year I was about three or four hundred pounds off of my allotment on this farm. On my other farm, my husband's farm, they made up a lot that we had lost the year before. They still didn't come up to their allotment so they got to add on this year. It is all worked through the ASC office. I got a form from them last week what I am allowed for poundage for this coming year."

The farmers we interviewed find the present system to work well. Oscar Richards put it this way: "This program we've got right now is one of the best and the soundest programs that we've had. You've got your base, your basic pounds, on the farm. If you fail this year, you can still go back and raise that next year. Say you got ten thousand pounds this year, your basic quota. Well, if you don't raise that, next year you can raise twenty thousand pounds. Now that's good. That's really a good program."

There are two important constraints in the carry-over provisions. Regulations do not allow the farmer to accumulate carry-over beyond 100 percent of the quota. That is, if the farmer's basic marketing quota is ten thousand pounds, he or she could not accumulate carry-over beyond twenty thousand. A more practical limitation is barn space for curing and labor for housing. Typically most farmers have carry-over in a given crop year; therefore, the barn facilities in a locality get swamped.

Farmers can lease out their quota to another farmer. Then, the tobacco is grown on another farm and they are paid for use of their poundage allotment. Lease prices are highly variable. When there is widespread underproduction and many do not make their quota, the price is low. When the crop is good and many are making their quota and some excess, the price goes up. Prices are higher in the marketing season generally unless there is a very short crop. When the crop is good, marketing season leases occasionally go as high as half of the net selling price. Lease prices are higher where tobacco is a more important portion of farm income and where there is less alternative employment.

A farmer who overproduces may also lease in poundage from someone who raised less than was allowed him. Jimmy Bridges gave an example of this, "Last year, for instance, we raised too many pounds and we had to lease in some pounds from some other farmers. They didn't raise their quota, so they sold their pounds for fifty cents a pound, and we used that to sell our excess tobacco." The ASCS office aids this process by providing sign-up sheets for those who wish to lease poundage in or out and recording the leases. When this happens, the producer's allotment is temporarily changed to reflect the number of pounds leased in.

In the early 1980s, various sting operations were organized to increase farmer compliance with the provisions of the regulations for leasing. In these stings, U.S. Department of Agriculture staff posed as farmers. At warehouses, they told people they had not filled their quotas, offering to lend farmers their marketing cards to sell excess tobacco. This produced a number of guilty pleas in the federal court in Lexington.

In 1991, marketing regulations were changed to allow the sale of tobacco quotas independent of the land. The intent of this was to increase burley production so as to help the United States maintain market share worldwide (Osbourn 1991, B4). Quota sales are subject to a number of restrictions. Sales can only occur within the same county. Sales of quota for a given crop year can occur up until the first of July and a person can buy up to 30 percent of his or her quota or twenty thousand pounds whichever is greater (Osbourn 1991, B4). Initial indications suggest that sale prices vary with lease prices, most buyers are larger producers, and most sellers are small producers.

Since the implementation of the poundage system, there has been more or less continual short production. Tighter supplies have made manufacturers less able to be selective. They ended up needing most of what was produced, and they paid almost as much for lower grade tobaccos as higher grade tobaccos. This has resulted in the reduction in the incentive to carefully grade tobacco.

The fate of the tobacco program is a matter of national debate. Ideally, most of the tobacco marketed each year is bought by the companies, leaving little for the pool (i.e., the Burley Tobacco Growers Cooperative). What is acquired by the co-op is processed and later resold.

The high price of domestic tobacco and the increased quality of foreign imports have led to increases in the market share of imported tobacco (Snell 1991). As a result, when harvests are good, tobacco stocks accumulate in the pool. Excellent harvests in 1981, 1982, and 1984 led to large surpluses in storage. For example, in 1984 30 percent of the crop went to the pool. This situation was compounded by the disastrous drought of 1983, when 50 percent of the crop went to the pool because it was of undesirable quality (USDA 1981, 1982, 1984).

Increases in the amount of tobacco in the pool increased program costs. This provided justification for increased attacks on the tobacco program by the growing number of persons concerned about the health effects of smoking. In the context of the changing political meaning of the tobacco program, Congress passed the No-net-cost-program Act of 1982 which "mandated that the U.S. Tobacco Program operate at no expense to U.S. taxpayers (Snell 1996)." The costs of the price support program were funded through a fee assessed on each pound sold. These no-net-cost fees are paid by producers and purchasers when the tobacco is sold at the warehouse. Some administrative expenses of the program are paid from taxes.

In 1985, the Tobacco Program Improvement Act was passed to deal with a number of very difficult problems (Smiley et al. 1986). These problems included high support prices that had reduced the competitiveness of American burley on the world market, tobacco pool stocks at record high levels, large increases in no-net-cost assessments, declining U.S. burley exports, and increases in the amount of burley imported for domestic use (Snell et al. 1991, 1). Although only 15 percent of the crop went to the pool in 1985, increasing production costs and no-net-cost fees left the farmer with declining profits. The Tobacco Improvement Act provided a number of mechanisms for managing these problems. These included lowering the support price level for burley, reducing quotas, cost sharing of the no-net-cost fee between grower and purchasers, and mechanism for selling burley stocks at a discount (Snell et al. 1991, 5).

Several farmers expressed doubts as to whether the support prices could remain so high without ruining the market. Oscar Richards said, "Now, the price program is just fine for tobacco. This is one of the government projects that has worked because they control both ends of it. They control the amount that you grow, which has to be done before they can put a price on it. That's what got them in trouble with this milk. They only had ahold of one end of it, the price support, and the government got stuck with a whole lot of extra milk and cheese and they couldn't do nothing with it, only just give it away. But now, on this tobacco, they got ahold of both ends of it and we've got a wonderful program, but they're going to break it down. They're pricing [us] right out of the market. The tobacco's too high. Not according to everything else, but it's too high, and if they keep raising the support price and cutting down on the production, they're going to get tobacco from overseas. Now, it is not as good tobacco as ours, but they are going over there and getting it and they can make it work. Now, they can change the blend of this tobacco and use that tobacco over there. It may not be half as good, but who's going to know the difference? You'll get adapted to the taste just like they did the cigarettes. You'll get hooked on it, and if they'll just hold that price down and let us raise the amount of tobacco that the manufacturers need, they're trying to cut the production down to where the companies is not going to get near what they need. And they are going to be bidding against each other to get it, and that is going to ruin the program."

5
Sowing the Beds

The process of raising a crop of tobacco starts with the preparations for producing seedlings for transplantation. Although transplant production technology is changing rapidly, today the majority of tobacco plants are raised by the farmer in a nursery bed located adjacent to the field and transplanted. These seedbeds are prepared in the fall or spring and are often placed on the most fertile soil. Although not a recommended practice, the same location is often used over and over again, supplemented by heavy applications of fertilizers. Beds are plowed and carefully prepared so that the soil has a fine texture. Then they are treated to reduce weeds and other pests. When the weather warms in March or early April, the beds are sown with tobacco seed and covered with a fabric covering that serves to increase the temperature of the beds and protect the developing plants. This system is declining in use. Approximately 40 percent of tobacco plants produced in 1993 were produced using the float plant system instead. In this method, seed is planted in multi-celled, polystyrene trays filled with a planting medium. These are floated in a water bed in a heated, poly-film "glazed" greenhouse and then sold to farmers for planting. This system is discussed in Chapter 6.

Currently people treat beds with a toxic gas to kill weeds. Before this technology developed, farmers used heat to sterilize beds. Beds were burned in various ways. Some farmers heaped the entire bed with brushy materials and set it afire. Others burned beds section by section. Earlier, farmers used large logs that had been set ablaze and then manipulated them with long hooked poles. This hot and demanding work could take all day.

The wood used for burning beds was often cut during the winter and spring, after other chores were finished, and then allowed to season. Some farmers got a year ahead in their wood cutting. Farmers who had fields on larger rivers might use driftwood. Others, breaking new ground, would use firewood that had been cleared off the land. In earlier days, more wood was available. The availability of chain saws in the early 1950s had an impact on the process, although, as Lawrence Simpson of Fayette County quipped, "The chain-saw era didn't last long" because of the substitution of chemicals. In the 1950s, some farmers began to use sawmill slabs from the mountains. After experimenting with this fuel, farmers found it worked better if they added some green wood to slow down the fire.

Burning of beds got things cleaned up on the farm, according to Christine Sims of Robertson County. "[Burning] took care of a lot of debris that collects nowadays. It kept the country much cleaner because the dead trees were used in burning on those plant beds. My husband did the burning by himself. He might have had some help sometimes. You hauled that lumber, dead trees, and anything that would burn and piled it. And, of course, he knew about what depth to pile the wood on. [He set it up] so that it would start at one end, and you gradually burned from one end to another. That left a deposit of several inches of ashes on the bed. And then after that, he harrowed that in the plowed up bed. It made a good base for the plants, you know, because it wasn't packed down."

Some farmers used horses to drag brush and other timber to the bed for burning. A retired Harrison County farmer, Allen J. Whalen, recalled his practice. "We burned beds. I've never gassed beds. The last crop of tobacco I raised I burned them with wood. When my dad got ready to burn beds in the spring, he'd make a [drag] out of brush. He'd cut down brush and cross it to make a hitch. When he got ready to burn his bed, he'd hook the horses to that pile of brush and drag it so that it would completely cover the bed. He'd catch the wind on a day when the wind was blowing down the bed, set the thing afire and let it burn the length of the bed. He'd take a forked pole and roll that brush from one end of that bed to another. What didn't burn he'd keep bunched up. He'd burn his britches and [burn] his eyebrows off."

Burning required that you wait some time before the seed were sown. Clara Garrison of Bourbon County said, "We would have to wait until late the next day to make sure that there wasn't no fire or warm spots to burn your seed. You know, little tobacco seed, wouldn't take much to burn it."

Burning beds created fond childhood memories. Berle Clay reflected on how enjoyable burning beds was during his childhood. "They were plowed in the fall and burned in the winter, and that was a fun event because you only started in the day. You'd pile up all the junk wood around the farm, and then you lit this fire on a sort of a skid built out of fence, and you would pull it down the bed, and it generally took all day long and well into the night. So you would be burning it in the middle of the night and that was like you had a big bonfire going.... Going out to see the beds being burned at night was a seasonal treat. [It would be done] any time during the winter probably as late as March. If you got the weather open and got it dry, you'd be burning the beds. [To pull the fire along] you made a sort of a sled. You generally have a long iron pipe which would be right angles to the bed and a sled, a wire fence laid out behind it. And you'd just lay the wood on top of that. I remember people experimenting with burning rubber tires which made an awful lot of smoke. I seem to remember that didn't work too well. The best thing was to get a good bed of hardwood coals going. Before this, there was pressure on picking up all the downed trees on the farm. That's something we don't have. Trees still come down but there's no pressure to pick it up. But then there was a real demand for blown down limbs or whatever to burn tobacco beds with. That [burning] killed your weed seed primarily. That's the main thing you are doing is killing the weed seed, as far as I know. I don't think you worried about the insects but mainly the weed seed."

More recently, farmers burned beds on a drag. In 1981 one of the authors assisted a farmer in Robertson County in burning a bed in this way. The drag consisted of a frame constructed out of welded pipe and pig wire and was as wide as the twelve-foot tobacco bed and seven feet deep. The large logs, brush, and old tobacco stalks that served as fuel were placed on top of the pig wire drag. He let it burn for ten to fifteen minutes on one spot; then using his tractor, he would drag the frame to the next section and let it burn some more. The drag itself was supported by a log that acted as a roller and soon burned up. As we made progress, the area that had been burned was marked by a pitch fork stuck alongside the bed.

Burning was relatively ineffective for controlling weeds, Reynolds Bell said. "We thought that burning killed the weed seed, which I question. I don't think it did, but the ashes was a tremendous soil conditioner. It furnished certain fertilizing agents, potash, and mellowed the soil." Some recall that burned beds were weedy beds. Arthur D. Jones of Bath County recalled, "Sometimes the beds would be so weedy. They'd take so many people to weed a bed. I have known weeds to take a bed. They would have to give some of them up once in a while. But you don't have a tenth of the weeds now. Most of the time [now] you seldom have to weed them. Then you would have to weed them day-in and day-out because the burning didn't kill the weeds like the chemicals does now."

Some farmers used steam to treat the tobacco beds. The steam was applied through the use of a specially built steam pan. Paul Carraco described this process. "It was a slow method. We had an old Huber return-flue boiler that we mounted on a truck chassis, and the steam from that would drive the old truck forward. First we'd get our ground prepared, and then we'd figure on getting the steam into the ground. Our beds were always ten feet wide, and you can make [the steam pans] whatever length you can get them through the field [to] a good spot to produce plants. These pans, we would make them up [out of] a two-by-four sheet metal and have them all where they wouldn't leak out any, except your edges. These pans would have a maybe a three-inch piece of metal bolted around the bottom of the two-by-four that the pans was up on, and this metal would settle down in the ground and keep the

A steam-powered J.I. Case tractor used for steam-treating tobacco beds in Fayette County during the 1940s. The steam pan has a wheeled jack to make movement down the plant bed easier. The edge of the pan is sealed with soil. (J. Winston Coleman Kentuckiana Collection, Transylvania University Library)

steam from coming out underneath of it. You'd put a little trench of dirt around the edge of this pan. These pans were nine by ten, and we'd run the ten length crossways which would make it the same width as the bed. If you had your pans made for ten feet wide, then that was all you could do. Steaming was quite a long, drawn-out thing. You'd have to make two pans, and you'd have this big steam engine and you'd have that steam just popping off practically all the time. It was hard on a steam engine to fire as hard as you had to fire to keep the steam up and blowing into the pans."

The system used by Carraco required two steam pans. "You'd have your steam hose running off from the side of the steam engine and then have a valve to turn on one pan at a time or turn them off. If you want to run one a little bit longer than the other, you could do that, or if you wanted to steam a little bit harder, you could turn off one and leave the other one on. You had [to] turn them both off when you stopped to move up another twenty feet. You'd have to have four people there to move those pans. It took a whole lot of labor. You'd set there and run that steam in there live for thirty minutes. You can see over a ten-hour day you couldn't get too far because it'd take you five or ten minutes to move those pans each time between your sets. And so when an hour was up, why, you didn't have quite two sets."

Currently, beds are gassed to kill weed seeds. Farmers cover the beds with plastic sheeting and seal the edges with soil. They "pump" methyl bromide gas under the covers and leave it on the beds for about three days. Some farmers gas their beds in the fall, some in the spring. The gas is purchased in pressurized cans, the contents of which are released through a pan into a tube which is placed under the plastic sheet of the prepared bed. The pan, equipped with spikes that perforate the seal of the pressurized can, is placed in a hole in the ground and mounded with earth to prevent tipping over. If the bed is on a grade, the gas is released up slope because it is heavy and flows downhill.

The bed had to be carefully prepared for gassing. As Paul Carraco describes, "We put this gas on in the fall. You have to have the temperature up above fifty degrees though, to use this gas effectively in killing weed seed. To use that gas, we have to work [the soil] down. You have the beds all ready and dig a little trench on the outside of [the bed and] you put down the poles along side [and cover it with plastic]. You have to take a shovel and pour that dirt back around there and down in that little hole that you dug beside the pole. You put the pole in there to hold the plastic up off of the ground so the gas can get distributed underneath of it. It's a rather hard thing to figure out just which way is the best. What'll be best for one person this year won't be best for him next year. You just can't hardly come up with something definite that'll apply each year and that'll be good. Dry ground, wet ground. Some of them works good in dry ground, some of them works good in wet ground. You just never know which one's going to be the best for what particular year you're having."

Some people do seedbed preparation on a custom basis. The number of people who did custom work declined when a law was passed requiring a license to do this work commercially, although it was possible to do a few jobs for neighbors without having a license.

Gassing a bed is easy and effective but has some disadvantages. It requires a cash outlay and does not result in a bed of ashes. The bed of ashes improves soil structure, which eases the task of pulling plants from the bed and leaves the plants with a better root system. Some farmers thought ashes were necessary for proper bed preparation but learned ashes were not needed when they started gassing beds.

Seed technology has undergone significant change. In the past, people produced their own tobacco seed by choosing those plants that displayed desired characteristics. The plants were left to stand in the field until seed was ripened and the plants were somewhat dry. These were cut and hung in the barn or stripping room until completely dry. The dry tobacco blossoms were then lightly rubbed to release the seed. Care was taken so as to not include too much debris in the seed. Paul Collins recollected the process of seed processing at home. "When you'd let these tops bloom out, they'd have seed in them naturally. I seen [my dad] take the best plants and take those tops and tie them up high [together] in the stripping room. And then in the spring of the year when it came time to plant the tobacco, why, he'd get a newspaper and spread it all out on the table and he'd shell. He had his glasses on. And if any of the seed did not look good, he threw that away. But if it was good-looking seed, he put it in there."

On-farm seed production involved systematic selection. Ira Massie spoke of his experiences as a boy. "We would save three, four, five plants. We'd cut the seed head off just before frost. Take them in and turn them upside down and hang them up in the crib and let them dry. Then, sometime over the Christmas holidays usually, we'd take those seed heads and mash them out. Blow the chaff out and we'd have seed." Ira Massie described the approach used for seed selection when he was working with his dad. "We made quite a science of selecting the type of plant we wanted [for] the following year. We looked for uniformity. We looked for size. We looked for number of leaves per plant. We looked for how well those leaves filled out back on the internodes, you know, so we could get that yield. In those days, we were getting all of one thousand to eleven, twelve hundred pounds to the acre. And the only way we could increase our poundage was to make sure that our variety that we saved was the best that we knew how to select.

Some local varieties were sold under the names of Judy's Pride, Warner, Kelly's, Barnett Special, White Pepper, and Yellow Mammoth. W.B. Collins of Mason County, experienced both as a farmer and Extension agent, commented that, "the first varieties were developed by farmers themselves. Farmers saved their own seed. Tobacco is a self-pollinating plant: the pollen from an individual plant also fertilizes that plant. So a fellow would select a variety of tobacco, and [the plants would] become fairly uniform. Then after ten or fifteen years, twenty years, they'd have varieties named for the farmer who developed the tobacco." These seeds were never certified. Ira Massie said, "There was no official thing about this. It's just a man decided, 'Hey, I'm going to save a half acre of seed and sell those seed next year,' and that's what he did."

The University of Kentucky tobacco breeding programs developed primarily because of problems of disease and the need for disease resistance in tobacco plants. Kentucky 5 seed, resistant to black root rot, was introduced in 1930 (Smith 1981, 227). This seed was derived from burley being crossed with resistant Turkish and cigar tobaccos (Smith 1981, 227).

A retired agricultural Extension agent, James D. Wells recalled some of the effects of the development of disease-resistant seed. "The disease that probably goes back the longest of any that I am familiar with was black root rot. [There were others] in those early days: fusarium wilt, mosaic. We had no resistance to those. So we had to try to manipulate the growing procedure to overcome those. For instance, black root rot: today we don't worry about it because all of our varieties that are released now are resistant to black root rot. Back in those days, what you did [was] let the soil get a little more acid because black root rot will not grow too well in acid soil. Today some of our older farmers are carrying over from that because they are reluctant to lime tobacco ground to say a pH of 6.4 or 6.5 which we now recommend. Back in those days, you'd never recommend to a farmer to lime his tobacco field that high because if he did, then the black root rot would take [his] profits."

There was also concern for yield. One of the earliest widely used improved seeds was Kentucky 16, introduced in 1934. This type combined the resistant qualities of Kentucky 5 with improvements in size and quality. Carraco reflected on the use of improved varieties in the 1936 crop year. "I remember when I first went to work for Mr. Barker, he had forty-seven acres that year of tobacco, and out of that we had forty-six thousand pounds. That was just no good. And you'd go out in a field and maybe we'd have a ten-acre field and have it in tobacco, and maybe on one side of the field you'd have some plants would be up three or four feet high, and then you'd have another one right beside it that wasn't any bigger than it was the day it was set out. This was something that we had to contend with to get us a better crop. So then a year or two, well, the next year in fact—I'm talking about 1937 now—and the next year they had developed what they called 16. It was a pretty nice leaf tobacco. Tall, [it would] grow with twenty-four, twenty-five leaves on it, and we'd been used to, say, twelve or fourteen leaves, or maybe ten. That was the first numbered tobacco that I remember. This 16 came right along the year after I was talking about raising that forty-six thousand pounds on forty-seven acres. And the next year we had up to fifteen, sixteen hundred pounds average on pretty near as much acreage."

Today people buy their seed from seed companies. It is sold in packages, called papers, containing one level teaspoon for hybrid seed types and two teaspoons for standard varieties. Commercial seed production is a meticulous process. The F.W. Richard Seed Company of Winchester is an important producer of tobacco seed. While its production technique has begun to make exclusive use of the float plant system, previously the company maintained the

B. L. KELLEY & SONS
TOBACCO SEED

Do you want tobacco seed grown by a practical farmer, on his own farm, hand picked, hand cleaned and with every care taken to insure highest possible germination? Do you want seed absolutely free from disease, which produces the tobacco that commands the best prices on every market, and has done so for almost half a century? If you do, get

B. L. Kelley & Sons "Improved Standing Up Burley" Tobacco Seed

None genuine unless in sealed packages bearing their registered trade mark, which is their facsimile signature.

B. L. Kelley & Sons
Lancaster
Ky

Price, $1.50 an ounce; ½-ounce packages, 75c; ¼-ounce packages, 50c.

ADDRESS ALL ORDERS TO

B. L. KELLEY & SONS, Lancaster, Ky.

A display ad for tobacco seed from a central Kentucky supplier published in the 1920s. (University of Kentucky, College of Agriculture, Library)

same seedbeds for each type from year to year to help protect the purity of the strain. The beds were gassed, just like the regular production mode, but gassing also has the intention of killing whatever seed from last year is viable in the bed. If any seed comes up from the previous year that was not killed by the gassing, it will be of the same type as is sown for the current year because the same beds are used.

The seed plants are transplanted beginning May 15th. F.W. Richard cultivates and fertilizes in a way that is the same as other producers, although it sprays against insects more aggressively.

During the first two weeks of July, the company evaluates and culls plants that are not true to type or are defective in other ways. This is called "rouging" and is done two or three times. About this time they get equipment together for pollination of hybrid seed although not all of the seed they produce is hybrid. They take great care to make sure that the pollen is of the correct type.

F. W. Richard described how the company organizes the pollination work. "We use boys and girls, but we don't let them work in the same fields. That's very important because of the fact that it just causes confusion. We let the girls do the pollinating. The boys pick the pollen out of the field. We bring it to the warehouse and dry the pollen down 'til it's powder. The girls put that powder in each and every bloom on that plant in order to have any seed." The process of producing hybrid seed is much more labor intensive than non-hybrid seed, consequently hybrid costs twice as much as the regular seed. Seed pods are harvested by hand because they mature at different times.

Since the development of a tobacco seed industry, the farmer must decide which tobacco variety to plant each year. A preferred variety may be sown for several years if successful. If the farmer has attended programs sponsored by the UK Agricultural Cooperative Extension Service, in which new varieties have been introduced, he or she may decide to change. Farmers also do their own informal experiments with varieties.

Some farmers may not switch to a new variety all at once. Instead, they try some first. If that does well, they plant more of it the following season.

Once the seeds are selected, the next step is sowing the beds. A rule of thumb used by some farmers is that tobacco beds must be sown by the 20th of April, with the sowing season starting in late February. Others we spoke with liked to sow their beds by April 10th. Many times the weather stays so wet that the farmers must wait later than this to get their beds started. Jimmy Bridges described sowing the beds, "In March and early April you sow your tobacco bed, you plow the farm, and maybe you sow about one bed for every two or three acres. [You use] about a couple of teaspoons of tobacco seed for a bed 100 feet long by 12 feet wide. You prepare the bed like you do an onion bed, just real smooth and nice. Then you fertilize it."

The sowing of this seed is accomplished in ways that have changed through time. In the past, the tiny seed was often mixed with sand, wood ashes, or corn meal in order to aid in even distribution on the bed. Now the seed is mixed with the fertilizer and both are broadcast. These days some farmers use small lawn fertilizer spreaders for this task. The fertilizer does not burn the plants because the watering and rain will incorporate it into the soil. If the seeds are sown too thick, the bed must be thinned out. Describing how it is done, Lucian Robinson said, "It would be right difficult to cover twelve hundred square feet [with the seed itself], so you mix [the seed] into [sifted] wood ashes. You usually line your bed out in about three strips. You'd have three strips four feet wide or you could mark one down through the middle. Since we don't have wood ashes too much anymore, you can use fertilizer." Clara Garrison of Bourbon County described a similar practice. "Dad done it with dirt. Most of the time Mama sifted in some wood ashes. We would have the wood stove in the living room. [Daddy] would tell Mama about a week or two in advance, 'Now you go to saving me some ashes and sift them so that they would be real soft to work with and save me,' he said, 'so many.' So Mama would do that, but if she didn't, he get him some fine dirt."

In the old days, beds may have tended to be overseeded. Ira Massie expressed his views on this problem. "If you look back in history, you'll find that nearly all of our beds were overseeded. Put too many seeds on them. [The plant population was] too thick. Our plants would be very poor at the time we got ready to transplant tobacco. [They had] very small stems and just did not have good liveability when we transferred them from [the bed]. I don't ever remember us pre-germinating them to see whether we had 70, 80, or 90 percent germination. Therefore we had difficulty in judging how many seeds to put on a bed. But that was standard in those days."

A teaspoon of seed would supply enough plants for an acre, according to Lucian Robinson of Robertson County. "If you plant

Tobacco seed. (Photographic Archives, University of Louisville)

Preparing a tobacco bed in Fayette County during the 1940s. (J. Winston Coleman Kentuckiana Collection, Transylvania University Library)

your tobacco in rows about thirty-eight inches wide and about eighteen apart in the row, you can plant on an acre of ground about ninety-eight hundred plants. [A] struck [i.e., level] teaspoon of seed is sufficient to sow a bed twelve feet wide and one hundred feet long, and if you have good luck with it and good germination, and weather factors are suitable, probably in the first time you pulled over in that bed, you could pull enough plants to set out an acre of ground."

Until recently it was common to plant some garden crops in the tobacco bed. Christine Sims noted, "We planted our lettuce around the edges of the tobacco. We grew onions, radishes, early things, you know. We would plant at the ends of the rows. So that was a good way to get some early vegetables before the regular garden was started."

Some farmers coordinated their planting with moon phases. This is called planting by the signs. Oscar Richards described this. "Some farmers believe in planting tobacco and other crops by signs. These are based on the waxing and waning of the moon. People used to try to sow their tobacco beds in a dark moon so they'd have better root systems. And then they'd want to set it in a light moon so the stalk'd go up. It works. Now, there's a whole lot in signs. A lot of these young people don't think so, but there sure is."

These days, beds are covered with a synthetic fabric for protection of the seedlings, although farmers may still call them "cottons" or "canvases." Farmers report that the synthetic fabrics retain heat better than cotton, accelerating plant growth and, therefore, allowing earlier setting. Since the late 1970s, seedbeds have been covered with synthetic fabric. The synthetics do not last as long as cottons, and some are designed for one season's use. The old cottons would be used for a number of seasons. Clara Garrison of Bourbon County recalled the care of these fabric pieces. "That cotton used to last for years. Mama used to wash that cotton out and hang it on the line. And we would use that cotton for years over and over." A very old farmer said someone told him that before cottons were available, they used cedar boughs to protect the beds.

Pulling plants from a Fayette County plant bed in the 1940s. The workers are seated on a low bench that reaches across the bed. The beds are covered with cottons. (J. Winston Coleman Kentuckiana Collection, Transylvania University Library)

Various strategies keep the fabric cover separated from the newly emerging plants. One approach is described by Paul Carraco. "Sometimes we put straw underneath that canvas. Just a real light layer of straw to hold the canvas up off the ground and keep it from rotting and keep it up where it'll be out of the way of the plants. Sometimes the little old plants will get started up through there and if the canvas is laying down right on the ground they'll come up between pores in your canvas. You might handle that canvas sometimes and break off those leaves that are on top of those little old plants. We have to protect them, and for that reason, we get [the covers] up off the ground just a little bit so the plants can go ahead and get started and get a pretty good sized leaf to shove up against that canvas then and shove it on up as it grows and gets long better." Farmers also attached the cotton to tree limbs all the way around the bed. This had the effect of raising the cotton off the ground by six to eight inches.

The cover captures some of the sun's heat during the cool spring days, which accelerates the growth of the plants. According to Paul Carraco, "When you put the cotton canvas over it, that adds a little heat to it. Keeps it where it will come on and grow real good once they get started and get out of the ground. But generally speaking, water is your limiting factor in starting plant beds for your tobacco crop, and when you were hauling it out on the hill land, and you wouldn't haul more than a barrel or two so that wouldn't amount to very much. It's become easier where water's more plentiful and we have irrigation systems. We can take [a] hose and water them this way, or we can put a tank on a truck and haul a thousand gallon at a time. People have fixed [it] that way where they can have a pressure pump and a thousand-gallon tank and just get on this truck and water it from the truck. It gives a good watering and keeps it coming on."

Other aspects of the management of the tobacco bed were described by Oscar Richards. "You have to keep the beds moist, got to water them pretty good. That is pretty important. And you got to keep the insects out and that is hard to do. So that is about the only basic thing there is to starting. And you got to set your plants at the right time, and as the old fellow says, 'When you go hunting, make sure your gun's loaded and the hammer back so you won't have to load it after the rabbit jumps up.' You got to be ready for things like that. You got to have your ground ready, fertilizer in, and then you got to keep your plants pulled 'til you get to set them out. If it is too wet to set them, you still got to pull them. After they get too big, you see, they runt down and runt the ones under them, so it is better to pull them if you have to throw them away than it is to leave them in the beds. You pull some of them, the biggest ones, and throw them away if you have to rather than to lose your beds. Now there was quite a few people this last year that lost all their plants. [Some farmers] plant back again, but I don't approve of that, and they just got plants from different people that had them."

If there is freezing weather after the plants are up, it may kill the germinating seeds or small seedlings. If the damage is great, the bed will be resown. If it is less damaged, the farmer will promote plant growth by watering and fertilizing.

Because farmers transplant during the last week in May or the first week in June, the seedlings are cared for for approximately six weeks. There can be a number of problems. In the days before gassing, plants often became infested with weeds. Christine Sims, speaking of the practices during the 1930s, reported weeding the beds. "And once or twice during the time that they were growing, they would have to be weeded. You put planks across the beds so that you wouldn't have to step on the bed. And you set or crawled or anyway that you could do it, on those planks and you weeded it. Pulled the weeds out on the side as far as you could reach. I could do that real well because I had the ability to double up and it didn't bother me." Edna Bell of Scott County noted that, "Then, you didn't put nearly as many chemicals and things on beds as you do now. Then, you didn't have anything to kill all those weeds. You had to weed the tobacco beds. Get in there with knives and get the weeds out."

Weeding beds was time consuming, as Arthur Harney Jr. recounted. "That was tedious work. I saw old farmers set on tobacco beds all spring, you know, [when] some of the younger sons or somebody were doing the plowing and harrowing and getting the ground ready and putting the fertilizer on. They'd have so many weeds that he'd set there on that tobacco day-in and day-out. That job I didn't like. Sometimes even with gassing beds, we still have failures and beds have to be weeded."

Through the years, there have been changes in the way plant bed problems have been handled. Blue mold is a problem that affects seedbeds and the fields. In the 1940s county agents emphasized daily removal of tobacco cottons for as much of the day as possible to toughen the plants to control blue mold. Farmers were to keep plants covered if it was raining. "Nitrating the bed" with nitrate of soda or ammonium nitrate in water was also recommended. These treatments were to be used only when the farmer heard about the problem in the neighborhood rather than routinely. In the 1980s, a chemical called Ridomil was used and was much more effective. Because this chemical was designed for use on the soil rather than on a standing crop, lettuce and radishes planted on the ends and margins of tobacco plant beds for household consumption were no longer safe to eat. In the past, there were restrictions on the use of plots upon which Ridomil was applied for forage crops in the following year, but this is no longer true.

After the seedbeds have been planted and the farmer is waiting for the seedlings to grow big enough to be transplanted, or set, in the fields many tasks must be done to prepare the fields to receive the young sprouts and in other areas of the production process. Paul Carraco described these processes. "We go ahead and plow it up, turn under the cover crop if we have a cover crop or turn under the sod if we have a sod, and drag it back down. Keep it worked down. And just do everything you need to keep it in good shape and good tilth so that things will go ahead and grow in it when we do get them in there. But it's quite a little time in here between now and setting time, so we have a good bit of things to think about and do. But you want to watch the beds then real close because it will get dry on you if we have a dry season. You want to have things ready to go whenever you do get ready. Then you keep your barns in repair. You've got to think about them, too. And if you had any tiers that broke down last fall when you were putting down tobacco, or whatever the problem might be, you want to keep it up and keep it in shape. There's always maybe some shingles blown off the barn, or blowing up a piece of tin roof. Oh, it's just so many different things that can happen to it. So you have to watch that and keep them ready all the time.

"New shares to put on a tractor plow, get it ready to plow because you generally always have to plow. Your weed killer you use may not get them all and so then you have to go over it with the plow. And then sometimes there are some big weeds that come on and get scattered around your field, and you have to go ahead and go over it and chop it. Some people don't like to do that and let it go, but it makes for better quality tobacco and it makes for more pounds to the acre, too, if you will go over and chop it and get those big weeds out instead of letting them get just as big as the tobacco because they certainly don't do the tobacco any good.

"When you get ready to set, we want to think of our sacks too. We use sacks [to] transport them from your plant beds to your field. [We] put these plants on these sacks real straight, and then pin them together with a nail.[1] So you want to have your sacks all ready and not have to go and get them the morning that you want to start setting them.

"There is so many different little things like that takes up so much time. When you put your weed killer on the tobacco ground, weed and grass killer, you can do that the same day that you set if you want to. Go ahead and get that done in the morning, and then pull plants and then set in the afternoon. You have to have your fertilizer in there, too. You have to be sure to put that on a few days ahead of time. You don't want to get the weed killers on too far ahead, but you can put fertilizer on two or three weeks or a month [ahead], but it's better to use it just about a week or a few days ahead of your setting. While you are waiting around getting a lot of these other things fixed, you can go ahead and put your water tank on your big truck or wagon, whatever you haul the water on. You can go ahead and get that done and get that ready to go."

The beds have to be cared for during the crop year after their annual crop of young plants have been pulled and transplanted. Paul Carraco described this activity. "After we've used our plant beds and set our tobacco out and are through with them, we'll go in there then and turn that plant bed over. Turn it underneath to rot, and then after we get our tobacco cut and in the barn, which is generally, like I say, September, we go and work those. We may put manure on those plant beds too, after we turn it over. [We] haul manure out of the barn or pens and run it on just as heavy as we can with one trip of the manure spreader over the plant bed. So this is something that you're going along with in July and August. And you're keeping those weeds down in those plant beds and keeping them trimmed down around the edge, keep the grass from going to seed and the weeds from going to seed and everything. You just want to keep it in a good state of tilth."

6
Setting the Plants

Tobacco is transplanted, or set, from mid May through early June. This involves either raising or purchasing transplants, preparing the fields, and transplanting using a machine (called a setter) derived from machines developed for planting vegetable seedlings. This aspect of the process is now undergoing a major transformation. Increasingly plants are raised commercially in greenhouse operations rather than being grown on the farm for use on that farm. In addition to changes in plant production, the planting itself has undergone substantial change during the lives of the older farmers with whom we spoke. There has been a shift from the use of the simplest hand tools to intricate tractor-drawn, mechanical setters.

There may be as many as eight or nine thousand plants in an acre of tobacco. The space between rows ranges from about forty-two to thirty-eight inches. Between plants, spacing ranges from twenty-four to eighteen inches. Currently the University of Kentucky is recommending sixty-five hundred to seventy-five hundred plants per acre. This is a much lower plant population than that of earlier practices. Wider spacing increases the yield per plant and the return per acre.

Before a farmer can set a tobacco crop, he or she must plow, harrow, and in other ways prepare the tobacco ground. In early days, horses and mules were used for field preparation. Reynolds Bell of Bourbon County described the process. "You used two animals to pull the breaking plow to turn the soil over. Then [you use] a disk harrow that's pulled by the mule. That cuts it up and stirs it. And that operation [is done] maybe a couple of times before [the] crop goes into it." According to Allen J. Whalen of Bourbon County, this early type of plow, "did a fairly good job but didn't cultivate the ground as deeply as it should have."

Plowing early is thought to have advantages under certain conditions. Christine Sims recollected plowing early. "I know that in 1930 [my husband] had plowed his ground in the early part of March, last part of February. Turned it over, you know, and it got some freezes on it and it stored moisture that was very useful because there was never any rain the rest of 1930. That crop was raised on the stored moisture from that early plowing. So it just depended on the kind of weather, whether you plowed early or like that or whether it rained too much and you had to plow later."

Preparing the soil on hillsides required a different approach, according to Arthur D. Jones. "[In the 1930s], we took a team of horses, hooked it to a plow. If the land was level enough, you would go around the field. [But most of them put their tobacco on the hillsides.] You took [your] plow out one way and throw the plow on the ground and then drag the plow back to the other end and set in and come back again. So you are just plowing half the time. After I was farming, I bought what was called a hillside plow. [Regular] plows would cut about twelve inches; it would cut about eleven. And you plowed one way and get to the end and then flip it over and you would turn around and plow right back. So you were plowing going and coming. These other plows, you would have to hold it with one hand and drag it all the way back and start again. You'd plow with that team, walk, walk, walk, and, of course, the same way with cultivating."

Field preparation can have an impact on soil moisture content. Roy Greene described his motivations for carefully preparing his Montgomery County fields. The fields are plowed during the last week of April and the first week of May and carefully harrowed. "After you plow the field, you should drag it down to conserve moisture. Drag it down, you pull a drag, some object to just smooth it down. It seals it so you don't lose your moisture." The soil will be worked differently depending on the farmer's predic-

Setting the Plants

tions of weather conditions in the coming season. As Oscar Richards said, "The most important part of it [is the weather]. If you can kind of predict the weather a little bit, in a dry year you plow your ground, you can work [it] down, down tight. If you are expecting a wet year, you don't want to do that, you want to leave it worked up. If you leave your ground worked up, it will catch more water but it will go down faster. If you drag your ground down, when it comes a rain it won't catch as much water, there will be more run-off, but it will make a seal on top of the ground and that crust will hold the moisture to the top of the ground where you need it most to start a crop, but in wet weather you don't want that. I did that last year. Another thing, in dry weather if you plow your ground early, in wintertime, let it go back together a little bit, now you will have a better crop. It will hold more moisture. But now this past year I plowed mine early at home and it held too much moisture, so ground that was plowed late grew better tobacco than the early plowed. So you never know."

In early days, all tobacco was set by hand with seedlings taken from the seedbeds between May 25th and Memorial Day. Frequently, if weather conditions are all right, setting would get done between the end of May and the middle of June. Setting by hand required the use of a homemade dibble called, a "peg," or by one's fingers. A piece of root with a crook in it was often selected for use as a peg so as to help form a more convenient hand grip. Because of the need to reset plants, farmers often still have an old peg around. In planting with a peg, farmers had to take care to cover the root system using their hands or foot. In setting by hand a person might first drop the plants where they were to be planted or have another person drop them.

The field was specially prepared for planting by hand. Called "laying off," this involved using a special single plate plow to create ridges of loose soil within which plants would be set. Speaking of his father's practices, Ira Massie said, "He'd get the field laid off into rows that were thirty-eight to forty inches wide. [He] would roll dirt up on a little ridge." Laying off had an aesthetic aspect, and straight rows were a matter of pride. As Massie said, "They bragged on it. [They'd] go to town on Saturday and talk about how good John's field looks because all the rows are just as straight as a bullet. They were proud of uniformity, too. Each plant [should] be of the same size. That was another bragging point. We bragged a lot more in the old days than we do now. Course, we don't have much time today to talk to each other. We're always on the go."

"Pegging it in" was also hard work. Paul Carraco described aspects of using a peg. "You take [the peg] in your hand and carry it to dig a hole. It generally packed the ground around [the hole] so much that it would dry out, and then a lot of times the tobacco would die because of that. Because you're making a hole where you'd roll the peg around and it would just pack the ground right in tight against it and then roots just didn't have enough power to get in there and go ahead and grow back out." This technology was retained for a long time on sloped lands and is still used when farmers have to replace plants that have died early in the growth process. Farmers who have done any amount of pegging talk about it being difficult. James O'Rourke said, "It was a terrible job. It was a back killer. I personally have always thought the two hardest jobs on the farm were setting tobacco by hand and cutting corn with a knife and putting it in a shock."

Because of the difficulty in carrying water for each plant, setting by hand required that the soil be moist, or, as farmers said, there had "to be a season." As Spurgeon Louderback of Robertson County described it, "In olden times, they didn't have tobacco setters. They just set it by hand, and they had to wait until a rain came to make a season. And then they'd go out and they'd have their ground prepared and they'd set the plants by hand." Ira Massie describes the conditions that made a season. "[It] had to rain the night before. The soil had to be in just the right condition before you could take plants and transfer them into the field. If it's too wet, then you ran into a lot of trouble, and if it's too dry, the plants wouldn't live. You had to have the soil just right."

Obviously, farmers had much less managerial control over planting in the past. Before the development of improved setters,

Setting the Plants 87

THE NEW PERFECTION PLANT SETTER

It plays the "devil" with the "weatherman" and shoots the risk clearout of sight; laughs at a "dry moon" and lends courage to the farmer who makes it his own.

And it does your transplanting so much easier and better that every plant lives and grows right off—"three weeks advantage," the farmers say.

Used extensively in all tobacco growing sections. Embodies many "improvements" not found in other planters. A very practical and inexpensive Planter that insures you against drought. You will like it.

If your dealer hasn't it, order direct.   Fully guaranteed

PRICE $6.50 POSTPAID

Kennedy-Galbraith Planter Co., GERMANTOWN KY.

A display advertisement for a hand setter manufactured in Kentucky. The plant is dropped down the tube on the left. The tube on the right is a water tank. This ad is from *The Burley Tobacco Grower* in the 1920s. (University of Kentucky, College of Agriculture, Library)

"everything had to be in harmony with the weather," said Christine Sims. "Your plans had to coincide with what the weather was going to be. When planting was done by hand, when it rained and you were ready to set your plants out, you planted with the season because it had rained enough so that you could."

Around 1925 an improved hand setter was marketed in central Kentucky. The Kennedy-Galbraith hand setter provided for the watering of the transplant. This allowed the farmer more flexibility in scheduling because they did not have to wait for a "season" as they had in the past. A Robertson County newspaper item from 1931 stated, "The farmer in times past waited for a season to pass along to set his tobacco, and in so doing, got left. He now buys one of Harry Galbraith's tobacco setters and sets his crop when his plants are ready."

The hand setter required at least two people to plant, one to drop plants and the other to operate the machine. In the late 1970s, some farmers reported that they used these hand setters for resetting where tobacco had died or washed away and for planting other transplants, such as beans and tomatoes.

The workings of these ingenious devices were simple enough, as Paul Carraco described. "A jobber was made up of metal, and it had two sides to it. It had one side where you'd have water; it would hold maybe two, two and half, three gallon of water. And the other side was a metal tube about six or seven inches broad and long, about six or seven inches in diameter. Your water would come down one side and you'd have a man going along beside of the [person doing the planting]. This man operated the setter with two hands, and he had to carry the water in it. And the other man dropped the plants in [the other chute], which was like a little downspout. And [it] had a nose on it down at the bottom, and a little handle up on the top where you'd squeeze this handle, and that would raise this nose apart, and then you would lift it off of the plant then. And you'd also give it a little squirt of water with the other hand. Then you had to take your foot and press it in the side a little bit to keep [the plant] from falling. This setter wouldn't pull the dirt back to it; it'd just shove it away where you could drop it into the little hole, and then you had to provide some means of getting it back. A setter soon got on how to do it, and he'd just take and kick a little dirt to the side and step right to the side of the plant. You had [to] press it down and retain the water in it. The tobacco did real well behind those one-man jobbers. Of course, it really took two men to run them because one had to drop the plants in it, but they'd get half an acre, three-quarters, up to an acre a day. Some people got pretty efficient with those, and even in times when we were beginning to grow tobacco in the bottoms, and then they still operated with these setters."

Setting with hand jobbers was time consuming. Ira Massie said, "You'd be setting tobacco for a good three weeks to a month anyway. You just didn't have the manpower to hurry it up and get it done any faster. [We'd] usually start [setting] about the same time we do now, May 20, May 25. And then set right up 'til almost

Setting the Plants

July 1. Today, [setting] twenty acres? It takes us about three days. Back in those days, five acres we'd set for a month."

Earlier in the century, some farmers used a horse-drawn "shoe-type" planter of the Bemis, Tiger, or New Idea brands. This implement consisted of a plow-like device that simply opened and then closed the furrow within which the plant was placed by hand. The shoe-type setter was not so hard on plants as the modern setter because you did not have to remove so much dirt or wring the tops of the plants off. It was much dirtier and more dangerous to work with because you were so low to the ground you could be injured by sticks and rocks. Your hands actually got in the dirt, and you couldn't see obstacles down the row.

The Bemis setter was much simpler mechanically than later transplanters. Silas Cleaver described how the Bemis setter worked. "You actually put the plant in the ground manually on a Bemis tobacco setter. It had what they called a knife and a shoe. A knife cut a slit in the ground and came back in a sort of a Y on the back and left you a small opening to place the plant in and then the dirt would close around it. And the shoes were flat with a curve like on the outer edges that would draw the soil in and firmly pack it around the plant. Ground had to be dry and loosely worked to do a very good job. You had to keep your shoes nice and clean and shiny for the dirt to slide in under them. In the old Bemis setter, you were stretched out behind the water tank and your legs stretched out flat and you set on a small pan seat which wasn't very comfortable and very dirty. You practically drug on the ground. The barrel set on the front of it, and if you were using horses, you sat up on the barrel. There was a seat on the barrel and two seats behind. And the droppers that ride it put the plant in, alternate every other plant. It had a water mechanism on it. The ground wheels powered this water mechanism, and it tripped a thing that would allow the water to come out of the barrel in approximately a half a pint or less per plant. You could adjust that. It would make a clicking noise and you were to put your plant in when it clicked and the other man that rode with you put his on the next click. And you had to time that right. If you got out of time and didn't put your plant in at the right time, then you, what

A display advertisement for an improved shoe-type setter manufactured by New Idea. Note how low to the ground the crew is. This machine is shown with an optional fertilizer applicator. This ad is from *The Burley Tobacco Grower* in the 1920s. (University of Kentucky, College of Agriculture, Library)

they say 'miss the water' and your plant would die. They made a Tiger setter which was just exactly like the Bemis design. And the New Idea made a setter which, instead of using the flat packing wheels, had little rolls that packed the soil around the plant. And the droppers were in front of the water barrel. As tractors came along, people stopped using horses and pulled them with tractors but it was the same piece of horse-drawn equipment."

The men on the planter had to accurately place the transplants, as Paul Collins described. "We didn't have a [setter] in the beginning. We were not too flush with money. Uncle Pat down on Elk Creek, my father's brother, he bought a setter, an ol' Bemis. We borrowed that. It was pulled by two horses. Setting in dry weather was quite a thing. It did pretty good. You had to be on the job to put these plants down in there when the water was coming. You see, the wheel turned to pull the valve to let out just enough water for the plant. You had to be down there ahead of it. Some fellow would get nervous and would get so they couldn't set. They tried too hard. I used to set ours. We got so we did pretty well. You'd change seats. You'd set with your right hand for two or three hours and then get over on the other side and set with your left hand."

The companies producing these transplanters worked to improve their designs. According to Bud Rankin of Bourbon County, the New Idea setter had significant advantages over the earlier Bemis and Tiger machines. "It came out with a different mechanism that packed the dirt around the plant. You could set a lot bigger plant. With the early setters, you had to pull your plants early when they were real small. With the New Idea setter you could let your plants grow in the bed and you'd have a big healthy plant, a stronger plant."

George Duncan provided his view of the shoe-type setter. "You take [the plant] and drop the roots down, of course, and there would be a little squirt of water come out, rigged up by gears all on the axle of your setter where the wheels were turning where it rode on. Of course, you had a lever on the front end that the driver operated, and he would raise this shoe out of the ground when you got to the end of the row and then when you got lined up and going back down the next row, ready to go back down the next row, he would raise this shoe up by a lever, and that would drop it back down to the ground again and the furrow would start being made. It would be a little trench-like furrow, about three inches wide or maybe something like that at the fullest part where you could put the root of the plant, leave the top of the plant sticking out of the ground and put the root of the plant down in this groove, and then this same shoe that came along then with a couple of sides to it, and this dirt was, of course, had to be worked up as loose—you didn't have any clods in your tobacco ground—and this would drive the dirt back around the roots then after the water had been dropped down in there, and it

A typical one-row wheel-type setter as manufactured by the Holland Transplanter Company. The water for planting is supplied by a drum mounted at the front of the tractor. The transplanter is mounted on the tractor and is lifted hydraulically during an end-of-row turnaround. The plants are in trays in front of each worker dropping plants. Missing plants are replaced by the person on foot. (Kentucky Department for Libraries and Archives)

just made a nice-looking job, and was a good job whereby plants would live. And I've said something about a lot of people would miss the water maybe when they dropped the plants down in there, dropped it too quick or not quick enough. And then if you continue to miss the water then you would have a bad stand. So a lot of people just turned the water on all the time and let it run a small stream all the time and thataway you're always in the water. It may not been as much as it would been if you'd hit it with a single squirt of water. But it worked out real good. It was very effective for us."

The modern "wheel-type" setter was developed in the 1920s for planting celery. The inventor, Ben Poll, worked for the Ford dealer in Hamilton, Michigan, and started to build machines for neighbors based on word-of-mouth marketing. Ray Kolk, a work

Setting the Plants

The general structure of a wheel planting mechanism can be seen clearly in this photograph of two women dropping plants. At the top of the circular path of the wheel, the plant holder opens to accept a plant and then closes. When at the bottom, the plant holder opens and releases the plant in the dirt. With shoe-type setters, the plant was held and released by the crew member. As a consequence, crew members were closer to ground level and actually had their hands in the dirt. (Photographic Archives, University of Louisville)

associate who became Poll's partner, took care of sales, contacts with dealers, and suppliers. These two partners split when the original patents expired in the 1950s resulting in two firms, Holland Transplanter and Mechanical Transplanter, both still dominant in the business.

Farmers regarded these setters as a significant improvement. Paul Carraco described wheel-type setters from his perspective as a farmer. "Setters progressed [to] where we came along with a setter that had a wheel, called a wheel setter. And this wheel had some little clamps on it, springs and clamps and rubber strips that held the tobacco plant in there and wouldn't mash it and hurt it. You [adjust them] anywhere from fourteen inches in a row to twenty-four inches in a row [by using] different cogs. We always aimed to keeps ours out around twenty or twenty-one, something like that. That gives you a little bit more room for the leaves to grow. If you get them too close, your leaves don't get long enough maybe because there wasn't fertility enough in there. You didn't have the fertilizer you needed in there. [These setters] had this water-dropping system. It wasn't like the first old two-horse setters we had. Those one-row setters were so nice and all that we never could get over having those developed to save that back and fingers and so forth that we used to get jabbed up when we were setting tobacco by hand. These setters have cogs in them that produce the water that turns around, and as the wheel turns and as the plant is dropped and that water squirts right in on the roots. So it's not a thing of where we used to have it, like we used to be where we didn't have the water to hit the roots. Maybe a man dropped it too fast or maybe he dropped it too slow or something [and] the water would be between the [plants]."

Farmers say plants should be set before June 20. If they are set later, the tobacco does not have time to mature before there is danger of a fall frost. Some older farmers placed a lot of importance on soil warmth. They thought that if they planted early, they needed to plant shallower because if they planted deep the cold soil would retard the plants. Planting depth also varied by soil type. Some older farmers planted deeper in sandy soils because the soil would heat up and burn the plants. The main reason for planting late is the weather. It is best to set the plants after a rain has softened the ground. However, if the soil is too muddy, farmers can't get in the fields with their machinery and setting can be delayed. If the ground is worked while wet, it becomes hard and the crop does not do well. Too little rain is also a problem. Some say the weather has more to do with the success of the tobacco crop than any other factor.

If the farmer is not using greenhouse transplants, the first step in transplantation is to "pull the plants." The covers are taken off the seedbeds. Some farmers set a rail on blocks over the bed to protect the plants while others work right in the bed on their knees. The workers selectively pull high quality seedlings of the right size from the bed. A healthy seedling is evenly green and should be between four and five inches tall. It should be planted after the root system is well developed but before the plant gets too big. The plants are pulled and are often bundled with burlap wraps containing about two hundred plants. Early in the process, it is as if the bed is being thinned. After plants are pulled, the bed comes back and can be pulled again as the remaining plants grow and fill the bed again.

When we did tobacco work, we found pulling plants very demanding because of having to work on your knees and take care to not damage the immature plants that may be pulled later. Nell Collins described pulling plants. "As soon as your plants get big enough, you can begin setting tobacco if your ground is ready. When [the seedlings] get up, four or five inches [tall], you pull them out of the bed and take them to the tobacco field where you're going to grow the big plants. It's hard to pull the plants out of the bed. You have to be careful not to step on the ones you don't want to pull. They got so they put a rail across to walk on."

As mentioned, rainy weather can delay setting because farmers cannot get their equipment in the field. Rain also causes the plant bed to grow rapidly, resulting in plants that are too tall. Large plants have a lower survival rate and do not fit in the planter. To deal with this, farmers wring the tops off the plants after they are pulled or cut them in the bed to fit the setter using a rotary string trimmer (like a Weed Eater) or even a rotary lawn mower. One person reported trimming the plants up to five times. The stalks get to be quite thick, which some farmers regard as a good thing. We have seen plants being planted that are little more than a stalk with a few tatters of leaves. The extent to which this can be done is limited, but the University of Kentucky has recommended practices for clipping plants to manage plant production. The practice can increase plant uniformity and stem diameters. Clipping can be started as soon as plant leaves start growing up off the ground. The debris from trimming with a mower can cause problems when it rots.

There has been significant change in how plant beds are managed, as Bud Rankin of Bourbon County noted. "Our conventional way of growing tobacco beds has changed tremendously because pulling was a terrible job. You had to pull them each individually, keep them straight. My father used to think that they all had to be same size. Well now you pull them by handfuls and put them on the setter. My father would turn over in his grave if he knew how we was pulling tobacco plants now. We even take a mower [mounted] on a tractor and clip those plants three times before we set them. Years ago people would say, 'You're ruining those tobacco plants.' [Mowing] evens them up. Slows the bigger plants up and lets the small ones come on up. Hit a rainy spell, you just take a mower and start clipping them." Clearly, plant production involves a number of complex management tasks involving field preparation and plant supply management.

There has always been movement of tobacco plants from one farm to another during transplanting season. According to Paul Carraco, certain farmers always seemed to be short of plants, either because of a bed management problem like disease or poor planning. "There's a lot of [farmers] that I think never did try to raise plants very much. They was always depending on your neighbors. I know we used to have some people that'd come around here just as regular as clockwork. They wanted to know if I had any extra plants or early plants. There was somebody all the time wanting [plants], maybe he had some coming on but they'd be a week later than his neighbor's, and so he's always coveted the neighbor's plants. If he could get enough plants from

Setting the Plants

Planting tobacco seed in float plant production. The equipment in the foreground uses a pneumatic system to pick up and then drop a single pelletized tobacco seed in the planting-medium-filled cells of a floatable Styrofoam tray. The men top off each cell with more planting medium. (University of Kentucky, College of Agriculture, Communications)

his neighbor, he'd go ahead and set it out and finish up and then just let his plants grow." The meaning of this has changed dramatically since the use of new greenhouse production techniques, which treats plants as a commodity to be sold. In older times it was more likely that a farmer would refrain from selling surplus plants, preferring to give them away because they "never knew when they were going to need plants themselves." One farmer recalled that the only time he ever sold plants was when the local judge, who had fined him $29.50 for some minor infraction, asked him if he would sell some plants. The judge asked the price and was told it was $29.50. He laughed and wrote out the check.

The greenhouse-raised tobacco transplants produced and marketed in central Kentucky eliminate the need for seedbed

Float plant production. Newly seeded plant trays are shown floating in the shallow tanks as men assemble part of the tank structure. The heating and ventilation equipment and plastic film glazing can also be seen. (University of Kentucky, College of Agriculture, Communications)

preparation, including gassing, sowing, and the difficult task of pulling plants. The transplants are supplied in a specially constructed polystyrene container that can be mounted on the planter. George Duncan, an agricultural engineer at the University of Kentucky, explained this development. "The vegetable industry looked at different ways of growing transplants in pots, trays, and greenhouses. Efforts were made to introduce [these practices] to tobacco starting in the '60s, but it just was not a system that the tobacco farmer cared to convert to. His method of scattering seed on a prepared bed, covering that with a cotton or later a synthetic cover, letting the plants germinate and grow, was the best, cheapest, easiest method until labor began to [be] more of a problem. Getting people who could pull enough plants to keep a transplanting crew going became more difficult. It would take four or five people to pull enough plants to keep two or three people

Setting the Plants

transplanting. Transplanting is a long, drawn-out process. In the late '80s some work on the use of greenhouse-grown plants was started in order to produce a plant that could be transplanted without having to pull plants out of a field bed. This technological breakthrough has really flourished in the last three years in Kentucky. [The float plant system makes use of] fourteen-by-twenty-seven-inch Styrofoam trays with two hundred plant cells in it floating in a waterbed with a plastic liner with five inches of water. The plants get their water from below and grow out very healthy. This was really a revolution in growing tobacco transplants. [The capacity to grow this way] is a unique characteristic of the tobacco plant. Its ability to grow that way simplified the growing of plants to where the average farmer could do it without high-tech greenhouse know-how."

This system has evolved considerably. The earliest systems, either in the fields or "right out beside the house," made use of very light covers and the water bed. From this beginning, farmers started to make use of greenhouses with heating and ventilating systems. This has allowed the use of faster planters adapted to float plants.

While the float plant system does save labor, its important advantage is that it allows the shifting of labor. Duncan describes this. "It shifts the labor to a different part of the season, so that it has some advantages of spreading labor out to a different part of the spring when the work loads are not as intense on the farm. It's still costing the farmer more money to produce the transplants [with the float system]. It has a management advantage: they are willing to pay more to produce plants that way than the conventional way. Float plants do not save much labor, they're not cheaper, but they offer flexibility, and maybe plant survivability and some other things are the advantages that do make it economically feasible. It's an easier way and they're paying to do it and are converting to it each year by large numbers." With the old system, plants were typically pulled in the morning for setting later in the day. With float plants, setting can be done in the morning and continue until dark if necessary. Float plants do not wilt after transplanting.

The propagation of float plants has encouraged changes in seed production technology. Recall that float plants are raised in Styrofoam trays filled with a special medium consisting of peat moss and other ingredients. Because of the difficulty in placing the very small tobacco seed in the trays, seed producers have developed pelletized seed. Using pill-coating technology from the pharmaceutical industry, this process results in each seed being coated with a water-soluble covering that is designed to dissolve or split when wet. This allows use of devices that can place a single seed in each of the two hundred or more cells in a typical tray at once.[1] While the devices used in large operations are driven with air compressors, plans are available for making gravity feed tray seeders and dibble boards on the farm (Miller, Hensley and Wilhoit 1994).

Another recent innovation is the production of primed seed, the effect of which is to shorten germination time and to allow germination over a greater range of temperatures than unprimed seed. To prime seed, the processor adds water to begin the germination process, but before the seed is actually sprouted, it is redried, suspending development. The advantage is that primed seed allow double-cropping in greenhouse production and lower costs.

The one-row tobacco setter, briefly discussed earlier, is drawn by tractor and requires two persons to load, or drop, plants in the planting mechanism. Dropping plants on a setter is much easier work than pulling plants, as you are seated comfortably, well out of the dirt, doing a task that may be monotonous but is comparatively easy and rhythmic. When working with pulled plants the droppers have a bundle or box of plants on a tray in front of them and their attention is directed at "the wheel" that slowly revolves between them. A complex of metal springs equipped with large special Y-shaped rubber bands are attached to the wheel. As the wheel rotates, the Y-shaped springs open to accept the plant placed there by the person dropping them. The wheel is linked to a valve that allows a metered amount of water (usually mixed with liquid fertilizer) to flow onto the roots of the plant to help the plant get started. With transplants, the system is similar but the

Setting the Plants

dropper has to place the plants in a rotary hopper and there are racks for the plant trays fitted. Other than that, the machine works much the same.

On turnarounds at the end of the rows, the planter is lifted hydraulically. If the water valve happens to be open at that moment, the dropper will turn it off. Periodically one of the plant droppers cleans debris from the "wedge" (a kind of chisel plow) that is mounted in front of the wheel. It is clear that plant preferences relate to the needs of the droppers. Droppers like good stiff stalks without a lot of leaf or dirt. All this is not the best for plant growth. As mentioned above, plant tops are wrung off. The set plants begin to wilt after a short time. By morning they are usually erect again. It is assumed that there will be resetting. The best conditions for plants after setting are low wind to prevent drying, no rain, and moderate sunlight and temperatures. A light rain after the plants are erect helps.

A period of time is necessary for the seedling to reestablish its root system after transplanting. During this time, the stalk and leaves do not grow because all the energy of the plant is being concentrated on the roots. The farmers say the plants are "getting holt" of the ground. If the seedlings are stunted by drought or disease, they are said to "runt down." This term is also used for sickly seedlings in the bed. Excess rain may cause the seedlings to "drown out" or "root out." If some of the transplanted seedlings die, they may be reset by hand. Resetting is difficult work. Mostly farmers just walk the rows replacing dead plants from a supply they carry in a bucket.

7
Cultivating and Topping

A field of healthy tobacco is a uniform height with long, broad leaves. In the growing stage, it is a uniform green color. Maintaining a healthy crop is important to the farmer because any flaws in the leaf may be reflected in the price received for the crop and in reduced yield. Weeds reduce yields. While the crop is "lying by," a farmer's concerns turn to protecting the growing crop from weeds and pests. He or she must watch carefully for the beginnings of pest or disease damage and weed infestations and be prepared to deal with them quickly. The technology available to protect crops in the past was quite different. This stage of the process has been transformed through the use of agrochemicals: herbicides, fungicides, and insecticides. As agrochemicals help the farmer manage the risk of crop loss, they also change the pattern and amount of work.

In the past, cultivation rather than agrochemicals was used to protect the crop from weeds. Cultivation itself was significantly influenced by mechanization of farming. The number of tractors on Kentucky farms increased starting in the 1920s. While it is true that tractors saved labor, they had other effects on tobacco production practices. For example, tractors reduced the length of time the standing crop could be cultivated because of the potential for leaf damage. An older farmer reported that before "tractorization" he cultivated tobacco until it was "up to the horses' bellies." Cultivation that late with a tractor damages the plants too much. Reynolds Bell noted that they did the heavy work with tractors but kept a couple of mules to plow tobacco for a while. In

Cultivating and Topping 103

A Fayette County farmer cultivating tobacco with a mule-drawn team in the 1950s. (University of Kentucky, Photo Archives)

some localities, horses and mules were used for traction until the late 1950s.

Animals used in cultivation became very careful about damaging plants. Paul Collins reflected on the use of these work animals in the days before tractors. "Back in those days we had a little five-shovel cultivator that one horse would pull. This little five-shovel had a small shovel, and you'd go twice in a row. We had a good line horse. You'd just talk to him and he'd walk just along. It's strange how those horses wouldn't step on that tobacco. They'd walk right up against it. You'd cover that row up and then come back the other way and cover that one up. Twice in each row, you did that if you set it by hand just as soon as it would dry off because it would get awfully hard if you didn't go in there and work it. When you set it with a machine, of course, it's dry already and it benefitted. After so long [the horses] would get so that they would be pretty good [at walking the line]." Ira Massie said his father's horses "were just as careful as they could be not to step on a tobacco plant because they knew what would happen if they did." He reported using horses coupled to a rastus plow. "A rastus plow was a two-plated plow in which the left plate would really go down deep and kindly stir up the soil. We called it 'plowing it hard.' We had to [plow] in those days because our fields were so dirty weed-wise. We'd plow three and four and five times just to keep the weeds down."

Horses and mules were different animals with which to deal. James O'Rourke, a Roman Catholic priest living in Georgetown, Kentucky, who sharecropped in Mason County, spoke of his view on the matter of horses versus mules. "More farmers had horses because, first of all, mules were more expensive, and, secondly, many farmers, including my father and most of us boys, were always a little afraid of mules. They were a little tricky. You never know for sure just when you got them exactly broke thoroughly. A horse, once he's been broke, you can trust him and walk behind him, but a mule might wait for the opportunity for five years to spite you or kick you. Or bite you. [Farmers felt this about mules] even though they're excellent work animals and they can stand a lot more heat than the horses could. Also they are very agile. You take a horse and just walk along pulling a harrow or wagon and [they may] just step right in a hole. A mule will see that [hole] and step automatically aside. Many farmers liked mules on that account: they would never get hurt."

With horses and mules, farmers had to raise corn and hay as feed. Shirley Wegner said, "You'd feed them corn three times a day. Keep them up of a night and feed them dry hay, they'd done more work than they would on grass. Grass wasn't no account to work a horse on. I done my own shoeing, shod horses all my life."

Hoeing by hand was also important during these days. James O'Rourke spoke of how hoeing was done. "Weeds would get a start and there was a lot of chopping and hoeing in those days. For example, the old-fashioned farmers believed, you know, they had to chop weeds. But even [when] the patch was clean, they believed that you had to hoe it up. Twice. That would take a long time, and if it hadn't have been for the farmer usually having a lot

Cultivating and Topping

of boys, that would have been an unending job. Now some of the brighter farmers didn't go in for all the hoeing. They used to kid and say there's no fertilizer in a hoe. A lot of the young men believed that it was just a method that their fathers used to keep their boys busy." A similar line of thinking was expressed by W.B. Collins of Mason County. "When I was a boy, you hoed every plant individually. And farmers thought for many years that hoeing was the most important thing you did 'cause you pulled a little dirt around each plant. . . . What you mainly did was cover up the weeds. But they didn't hardly realize that. They thought that hoeing [really did something.]" Ira Massie saw hoeing as a matter of custom. "Why we did it I don't know. We could see that tobacco really didn't need it, but [that's] what the custom was, and we're great people to follow custom. The custom was to hoe that tobacco, to pull dirt to it. My dad believed in that, having four boys around. About two to three weeks after that tobacco was out, we'd have to get out there with a hoe and chop all the weeds out of it and pull dirt to each plant."

Cultivating the tobacco was more important until herbicides came into more common use. Berle Clay of Bourbon County mentioned that, "up until probably 1970, no herbicides were used. So a very important thing was keeping the weeds down by plowing. Chopping weeds down by hand, that's a function that women could do, or would do, too. It's sort of a solitary task walking down a row of tobacco. It [was in] the '70s that you start incorporating the herbicides which we use now. If you get [the herbicides] in right and you have a good season, you don't have to do any plowing for weeds. You do have to do that plowing where you do what's called 'side-dressing' your tobacco. That's [done] after the plant has grown a little bit, and the plow throws the dirt up against it. But you're generally not plowing for weeds. So you're probably in the crop more than you are now. The chemicals cut down tremendously on the labor and the time involved in the crop. The first chemicals in use would have been in the late '50s when we suddenly realized Johnson grass was invading us. And I remember a man from Paris came out specifically to spray Johnson grass. He had a jeep with a tank in it, and he drove it across the farm to the spots of Johnson grass and he dribbled spray off of it by accident so he left this long winding trail of no grass all over the farm."

Today weed control efforts start with the application of pre-emergent herbicides when the fields are being prepared, but they also involve cultivation. Weed control continues when the tobacco is worked with a cultivator one week after setting. This process continues as the crop develops. Clara Garrison of Bourbon County made the point that you have to keep after the weeds in tobacco. "You would always plow the tobacco until it got too big to keep the weeds out because once you lay it by, the weeds could take you, you know. You have to just keep fighting them, that's what I've done this year. Chop weeds until I don't have no smarts." Fields are usually cultivated two or three times during the first five weeks after setting unless the season is extremely dry. If weather conditions are right, after five weeks tobacco is too large to cultivate without damaging the leaves. The practice of cultivating tobacco further along in the growing season was more frequent before tractors replaced horses. Like Mrs. Garrison, some farmers refer to this period between the last cultivation and harvest as "laying by."

Rainfall has a big impact on the crop during this period, according to Gary Palmer, a tobacco specialist at the University of Kentucky. "Many old-timers say it takes a dry June and a wet July to make tobacco. This is basically true because tobacco prefers a drier soil for initial root development. The worst situation is wet soil for the first four to six weeks, followed by an extended drought. Under wet soil conditions, tobacco will not develop much of a root system. I have seen tobacco that had been set for six weeks with no more roots than it had when set. Only after a few days of drier soil, new white roots could be detected."

If there was excess rain, some farmers would plow the ground to loosen the soil and let the water drain. Oscar Richards of Montgomery County described an example of this. "After we set it, when it got half a knee high, it started drowning out. The roots that was on the plant when it was set, they died, the feed roots and tap root died and drowned out and the plants started rooting out toward the top of the ground, and the plant kept

Cultivating and Topping

getting littler and littler until it got a better holt, and just two or three days of dry weather then takes all the moisture from the roots of it, for the roots are up here at the top and they just didn't get enough water. It set in dry then. Wasn't getting enough water to feed the plants, and just kept getting littler. After a while it kinder started growing a little bit, we got a good rain or two. And I thought it was going to do just fine, and it done got up ready to button, it come a rain or two then and it looked like it was going to grow right on and then in a few days it come about five or six inches of rain and it drowned it again. By that time it was too late, per' near too late to plow. I plowed mine but I damaged the leaves quite a bit. I had to plow it right quick, you know, and I plowed it through the mud, but it did help the tobacco. It loosened up the soil and let the water drain and it helped it a lot. And what we didn't plow, it didn't do anything. But that was the worst spring I ever saw. I have raised fifty crops of tobacco, that was the worst one I ever tried to put out. The ground stayed wetter, it was harder to get your plants set, and it was just a very difficult year from start to finish—one of the worst."

In addition to weeds, a number of pests can harm the tobacco crop. These pests include tobacco aphids (*Myzus persicae*), flea beetles (*Epitrix hirtipennis*), budworms (*Heliothis virescens*), thrips (*Frankliniella fusca*), tobacco worms (*Manduca sexta*), and grasshoppers (various *orthoptera* species). Today, pesticides, such as Diazinon, Sevin, disulfoton, and Orthene, are used to control these pests any time a problem is noted.

In the past, there was much less spraying for crop pests. Alex S. Miller of Millersburg noted, "You know, you spray for bugs and worms and everything you can think of now. With the spray you'd kill them. But those days, you didn't have that and you would set it out, and then you might have to go in there and you'd worm it by hand. Go by the plants, walk up and down the rows and pick off the worms. They'd eat holes in the leaves and everything. Nowadays, if you find many worms you'd go spray."

Of all the crop pests, tobacco worms produced the most memorable experiences. Many of the people we interviewed had childhood tobacco worm stories. Edna Bell recounted her experiences with tobacco worms. "When I was a child, they used to worm it by hand. Go through and pick [them off]. I remember when my brother Larry and I were quite small, we went with my mother and daddy to the tobacco patch. They were worming tobacco, and we were helping. And we put ours in a little old bucket and [took them] to the end of the field, and Daddy said, 'What did you do with your worms?' We had buried them and naturally they came right back up through all of that and went right back into the tobacco patch. You're supposed to kill them. The men snapped their heads off, but I never could. I had to stomp mine." Arthur Harney Jr. of Bourbon County spoke of picking tobacco worms as child's work. "When you're a little boy, [worming is] about your first farm work. Afternoons you could almost count on your dad saying, 'Well, boys, let's go worm tobacco.' And you'd start out, you might take one row and then as you got a little more experienced take two rows. We'd walk those tobacco patches and kill worms, and I mean we killed a lot of worms. In one row you might kill a hundred worms. Some of the worms would be as big as your thumb and some would be as small as just almost slivers, you know. You can see them when they're a half an inch long."

Silas Cleaver described highly toxic lead compounds that were used as insecticides. "They used a chemical called Paris green. It was put on by [a] blower that you carried through the row. [It] blew this green-looking powder on the tobacco. And I've heard them talk and speak of if it was going to kill you, look like it would, because we'd be covered up, just practically green with it."

Tobacco is subject to a number of disease problems as well. These can occur in the bed, in the field, and in the barn. Blue mold, rust, wildfire, black shank, brown spot, root rot, and tobacco mosaic are a few of these diseases.

Black shank is a fungus infection that causes tobacco to develop a black discoloration at the base of the stalk. Because the roots are afflicted, the plants become stunted. It can be spread in various ways. Sometimes tobacco planted near a creek will be infected. It may also be spread by watering the tobacco with infected water or walking from an infected field to a clean field. It can be prevented by careful placement of the fields or by rotating the crop every one or two years.

There were early treatments for black shank. Paul Carraco spoke of one such treatment used earlier. "The first treatment we had for black shank was Nabam.[1] It was a liquid, and we could put it on the ground with so much water, so much per acre or spot. You could just spot check it where we needed it. And if we'd have it in a field, we'd go out and pull up all the plants that we could see that were affected and then pour this Nabam solution on the ground and soak the ground. That would stop the black shank from spreading in a bigger area. Maybe you'd have three or four spots in a field where you'd have half a dozen plants, and that a way you could treat it and keep it down. Black shank was carried by the water out of the Kentucky River mostly when it first appeared here. It was hard to keep it from being spread. We would get real mad maybe at a neighbor sometime. He'd have a spot of black shank in his tobacco, and he'd be over there wading around through this black shank area and then walk over in our fields which is right across the road from it. We didn't mind him walking in our field, but we didn't like for him to walk out of his black shank-infested field over into our fields. You might make a good neighbor mad sometime by asking him not to walk through your fields after walking through his fields if you knew he had the black shank."

Today the most injurious disease of white burley may be blue mold (Atkinson et al. 1981). The tobacco plant contracts blue mold in the seedbed or in the field. If infested plants are set into the field, the disease causes serious damage to the crop. With blue mold, one observes spots on the top of the leaf; however, if you turn the leaf over you can observe the mold spores that look like blue dust. The University of Kentucky Extension Service recommends cultural practices to reduce the risk of blue mold infestation. The farmer should avoid plant bed sites with a history of the disease or where shady, moist conditions are present. They should locate their field in open, sunny areas and open up the spacing of their hills and rows. Ridomil is subject to restrictions in its use as a

Tobacco in blossom. (Photographic Archives, University of Louisville)

treatment to avoid contaminating the human food chain. Blue mold is now becoming resistant to Ridomil.

When the tobacco plant flowers, the "tops," consisting of a portion of the stalk and the apical flowers, are broken out. As Nell Collins said, "It blooms just like a flower, and those blooms have to be broken out, and they call that topping." George Duncan described the technical aspects of this. "Tobacco needs to be topped when it begins to bloom out so that the energy of the plant does not go into the bloom and seed, but goes into producing the leaves, which are marketed."

Some farmers begin to top their crop early when the plant "buttons" (i.e, in the growth stage, when buds are forming at the top of the stalk). Other farmers prefer to wait for their fields to come into full bloom. Topping practices vary in different counties

Tobacco being topped in the 1970s. (Kentucky Department for Libraries and Archives)

and neighborhoods. Alex S. Miller said, "Different farmers had different ideas on how low or how high they should top their tobacco. . . . I've seen people top tobacco a lot lower than the men that raised our tobacco did, and I've seen some of them top it higher. It was pretty hard to get these boys to break too much out if they thought those top leaves might grow and be more pounds when they got to market." Reynolds Bell approached the problem in a different way. "I always topped [early], the first time you see color in the bloom. Some folks let it bloom out, get big blooms. But I like to top early. That way the energy that would have been going to this bloom, seed head, is put into the leaves, what I was selling."

Generally farmers indicated that they preferred to top in the morning because it was easier to break the flower off cleanly. The break is down to where the larger leaves are located. This is thought to force the growth of the lugs and the brights (types of tobacco leaves explained further in Chapter 9). Some farmers top tall plants short and short plants long so as to even them out. During topping, insects might be picked off the plant.

"Suckers" are small growths that grow out from the base of the leaf stem. George Duncan described the plant's response to topping. "After you top tobacco it wants to sucker, like most any plant wants to try to perpetuate itself with suckers. Years ago the method of getting rid of the suckers was by hand, every plant. Go break out four or five suckers. And the longer the plant stood there, the bigger they would get and the quicker you'd have to get in to break them out." If allowed to grow, suckers reduce the growth and weight of the leaves on the main stalk. The suckers were broken out by hand and the process was repeated a number of times during the season. Now the plants are sprayed at the time of topping to inhibit the growth of suckers.

Suckering tobacco was hard work, as Eddie Reynolds described. "It was one whale of a job in those days to go through a crop of tobacco and break those suckers out of it by hand just before you cut it and put it in the barn. If they got too big and tough, you'd have to go through there with a hooked-beak knife and cut the suckers out of it. I remember during the war, when I was teaching school at Ruddells Mills, seeing a field with suckers on it long as your arm, and they're tough. And people had to cut those suckers out of that, and you cut it out and drop them on the ground. And that, I think, is probably the one job that's a lot better now with chemicals than it was then. Well, they'd just break them

Cultivating and Topping

A high-boy sprayer in the 1970s. (Kentucky Department for Libraries and Archives)

out and drop them down beside the plant, in what they call the bulk of the row. And they'd run over them as they were housing tobacco. And then they would go ahead and harrow them up when they got ready to seed the ground down."

Some farmers would leave some suckers on the tobacco in an attempt to manage the growth of others. Eugene Kiser of Bourbon County recounted his approach to suckering. "I'd leave two top suckers on [the tobacco]. And that would keep [the suckers] from going down to the bottom. Then when you got it pretty good shape and ready to cut, then you'd go and cut those two top ones. They'd be pretty good size. You'd have to cut them and then the tobacco would be ready to cut. Lot of times you'd get a piece of tobacco ready to cut, suckered off and everything, and it would rain four or five days and those suckers would go to the bottom of it. And you'd have to sucker it over again. And that's when you talk to yourself a whole lot. You'd go out in the morning when it was wringing wet and sucker tobacco. That was the time to do it because they broke easier and everything like that."

Suckering tobacco was more tedious than anything. "[Suckering] was a job a young man hated and an old man liked," said Arthur Harney Jr. "An old man could sucker tobacco and it wouldn't take as much energy as it would to cut tobacco or to hang tobacco in a barn or hand it up."

All farmers now spray their tobacco with a preparation called MH (maleic hydrazine). Originally developed by Uniroyal, it is sold under various brand names and called sucker dope, sucker medicine, or sucker cure by farmers. It was first tested for use in Kentucky in 1954 by University of Kentucky researchers C.E. Bortner and W.O. "Ted" Atkinson. Spraying for suckers saves a great deal of labor; estimates are as high as thirty hours per acre. Yields have increased because MH provides highly effective sucker control. Spraying with this substance accelerates the yellowing process that occurs prior to cutting. Bud Rankin suggested that farmers were cautious about the use of sucker control chemicals because they interfered with their ability to tell when tobacco was ripe for cutting. "I remember when it came out. Everybody was leery of using it. When they started to use it, some farmers believed you had to leave a row or two with the suckers and they would go by that row when they would decide to cut the rest of it. In the past, the crop was harvested when it turned yellow with maturity. One farmer implied that because of the effects of MH on leaf color, there was a tendency to use a "rule of thumb" of cutting twenty-one days after applying the chemical.

This chemical can be applied using a high-clearance self-propelled mechanized sprayer called a "high-boy" or a backpack sprayer. It is possible to hire a custom applicator to do the spraying. Spraying of MH can occur before or after topping, but most farmers top first because the MH toughens the stalk and makes topping more difficult. Weather is a consideration, too. The effectiveness of MH is reduced if it is washed off too quickly and the chemical can fall on the person doing the spraying in windy conditions. When we sprayed MH using a backpack sprayer, it was difficult not to get the chemical on your body.

MH works systemically. Fatty alcohols used as a contact spray are a substitute where there is concern for residual levels of MH or

Cultivating and Topping

if the MH is being applied late. Some farmers use a combination of both types.

Some problems occurred when farmers started using this chemical. Paul Carraco discussed the introduction of MH. "When we started using MH-30, the suckers would be controlled but you'd also have a lot of black tops where this MH-30 would hit on top leaves and it'd hit stronger there than any place else. We'd end up with maybe a lot of black-tipped leaves. People got onto that they'd get more pounds when they put the MH-30 on by putting on double amount, sort of like it was with the fertilizer. If they'd put on an extra gallon of MH-30 per acre or extra half a gallon, that would maybe add three hundred pounds to the weight they'd get per acre. Well, it got so that people were just so interested in pounds and all and paying very little attention to quality because pounds is what brought the money, they thought. And so when this thing happened that way, that made it kind of rough on our tobacco as far as quality was concerned."

Farmers recognized MH as a big labor saver. Edna Bell told of her experience with MH-30. "[Suckering] was a very distasteful job to everyone, so it was great when they got the MH-30 that they sprayed on it. I remember the first year we ever used MH-30. They didn't have any way to spray it, only by hand spraying. [My husband] just had a regular hand spray that you pump up. He fixed it so it had two nozzles on it so you could spray two rows at a time."

When MH was first available for use it was necessary to get a permit for it. As Vivian Owens recollected, "We had to go through [the Extension service] in order to get a card when they did start spraying tobacco for suckers."

After topping and spraying, the farmer waits three to four weeks before harvesting. This allows the plant to mature and ensures that the chemical is no longer on the plant when it is harvested. Chemical on the plant could present a health risk.

8
Cutting, Housing, and Curing

To harvest tobacco, a man cuts each individual stalk with a tomahawk-like tobacco knife and impales it on a tobacco stick. Cutting tobacco is arduous and dangerous work. Once cut, the tobacco is allowed to wilt in the field for a time and then is housed in a specially built barn and cured. Curing is a process of controlled drying in which the nutrients remaining in the leaf from photosynthesis are used to support the declining life processes of the leaf. During the curing time, the valuable leaf is subject to damage from too rapid drying or from the mold and mildew caused by moisture content that is too high. When the curing process starts, the barn is full of a raw agricultural commodity; when curing is complete, the barn holds a processed agricultural product. In other words, the initial stages of manufacturing occur on the farm.

Farmers debate the optimal time to cut tobacco as timing greatly affects the quality of the cured leaf. Tobacco cut too early has a high percentage of poor quality, immature, or green leaves. If the tobacco is cut too late, risks are leaf loss and reduction in quality and weight. Tobacco should be ready to house in ninety days from the time it is transplanted. When the field is ready for harvest, the leaf will have turned a uniform gold color. Oscar Richards of Montgomery County discussed his views on when to cut tobacco. "Me and Roy [Greene] has always disagreed on the time to cut tobacco. He claims if you cut tobacco green, it will cure yellow. Well, that is a sad mistake. I'll attest to that one year after another. I'd be out in the field when I was topping tobacco. I'd find me a

Cutting, Housing, and Curing

real good plant of tobacco. What I was trying for was to see what stage a plant should be in to make the most money, and I'd bring me in one while I was topping. Course, it was green. I'd find me a well-matured stalk of tobacco. I'd hang it up—now this was before we started spraying—I'd hang it up and go back in about a week or so. I'd go in there and I'd pull the suckers out of another good plant. Well, I'd take it and hang it up while it was still green but it would cure a little better than this other. Now the first one would cure red from one end of the stalk to the other. You could have just stripped it all in one hand. But the next one would have a few kind of bright leaves on the bottom and then from there up it would be red. And when it gets in the ripe stage that's when it kind of begins to kind of turn a little yellow in the top. Little yellow spots come to the top and just all over the top leaves. Now that is a good stage to cut it. It's got more weight and good quality, and it will make you more money. But he thinks when a stalk gets ripe, it will cure red. Well, that's when it has the high color, what he calls 'paw-paws.' But there is a lot of difference in paw-paw tobacco and high-color tobacco. An overripe stalk of tobacco that cures up kind of quick, it will be high color, but it is smooth color all over, the whole leaf will be smooth. And paw-paw tobacco is tobacco that grows in dry weather, it set in dry just a little bit too early before it matures. Then you put it in the barn. You take it off of a dry dirt, you put it in the barn, and it stays dry for a little while. It don't cure up as dry, though. It's got no moisture in the stalk when you put it in there and it takes moisture to cure a stalk of tobacco, and it will have them yellow spots in it, yellow spots and dark spots.

"If you have a good season and you do have a choice, now that certain stage there when it begins to turn yellow spotted in the top, now that's a good time to cut tobacco. But I always figure, as much as we raise now, if you don't have but just a little tobacco you can house it just the way you want it, you can put it in any time you want to, but the way we are, I figure if I can get a third of it just a little bit green, a third of it just ripe, and a third of it overripe, I'll blend that."

Because white burley tobacco is impaled on a stick for hanging tobacco in the barn for curing, tobacco sticks are placed alongside the rows of tobacco in anticipation of cutting. Farmers call this "dropping sticks." Some farmers use a high-boy sprayer, upon which a rack for the sticks is jury-rigged, to drop sticks. Paul Carraco described dropping sticks from a high-boy. "We use a high-boy to drop sticks with. Put a platform up on it, and two people rides up there with, say, six or eight hundred sticks on it. And then you can drop them off and leave them in the field for a week or two maybe. But you want to be sure that you have all the spraying done because if you go to putting the sticks in, then if you go through to spray, you're going to be breaking up your sticks. So you can't get too far ahead on some of these things and you have to kind of watch out and keep everything right along as easy as you can. And then sometimes you have a wet week that comes along. Maybe it rained on your sticks out there every day, and that'll get them soaked with water real good and make it where it won't help you get your tobacco starting to curing real good with all this moisture in the sticks where you put the stalk a-straddling the stick."

Today's tobacco sticks, produced by sawing, are not made on the farm. Earlier sticks were hand split with a froe and mallet or cut from saplings and made on the farm. Hobert Dooley of Rockcastle County remembered splitting red oak into tobacco sticks for sale. He later bought a special machine for sawing sticks. In older tobacco barns it may be possible to observe all three kinds of sticks. The modern sawn stick breaks easily and has no advantage other than availability. Sticks are sharpened on both ends and are about fifty-two inches long so they can be hung in the rails in the barn. Because the price of sticks has increased through the years, farmers lend each other sticks. This is done by counting sticks and bringing an equal number back. In this way the sticks in a neighborhood get mingled.

The current system of cutting was developed in the 1940s. Prior to this, farmers used a "shank" or "punch knife" to split the tobacco stalk from the top down, starting from the stub end where the stalk had been topped, down to about a foot above the ground. The stalk would then be cut at the ground. The split stalk would then be placed over the stick, straddle-fashion, about six stalks to a stick.

Cutting, Housing, and Curing

A shank knife consisted of a wooden handle, a steel rod, and a blade and could be purchased in stores. "It was a special knife," said Reynolds Bell. "It had a shank about fourteen inches long and a narrow-bladed knife about four inches wide and a wood handle on top, and you just push it down."

Although slow and more difficult, splitting had some advantages over techniques used later. Shirley Wegner recounted his experiences. "I started out when I was a boy splitting tobacco. And we split tobacco after that a right smart. It would cure up quicker when you split it. Open that stalk, let the sap out, and you'd strip, I'd say, months earlier than you would spearing it. It's just too slow and people couldn't make no headway of it. Six stalks on a stick. Now it was mean to hand up in the barn, that split tobacco was. When you'd hand it up that away, it'd slide off the stick awful bad on you. It's harder to split tobacco than it is to spear them."

The mode of cutting had an influence on how topping was done. In the old days farmers removed some of the top leaves with the blossom. According to James O'Rourke, "They believed that if a plant had twenty-five or -six leaves on it, [they would] break off the five or six at the top and leave about twenty good leaves. That would get [them] just as much weight and better quality." This also facilitated splitting the tobacco as the top leaves were more spread out. Today, as O'Rourke said, "the top leaves might still be growing toward the sky" when cut.

With this technique, you split the plant and then cut it. Paul Carraco recalled, "[You] take a knife, with a curved handle or a straight shank on it with a T-handle on it, and go to the top of this plant and split it down to within about eight or ten inches of the bottom of the plant. Then you would cut that plant off next to the ground, down below the last leaf, and you'd split it and lay it down. After you cut it off and let it wilt a little bit, you would go back with sticks, pick up these stalks, hang them on the sticks after you'd picked them up, and then lay the stick down. Maybe you'd want to put it in a pile so it wouldn't sunburn, or maybe you'd want to put it in a pile to kind of warm it up a little bit so it started to curing. And you had a whole lot of things like that you

Cutting tobacco with a tomahawk-type tobacco knife in the 1960s. The spear can be seen at the end of the stick. (University of Kentucky, College of Agriculture, Communications)

just didn't know why you did them sometimes, and sometimes you had to do them.

"Of course, in the old type of knife that we had where you'd split each stalk, you'd hit the nodes of a lot of leaves in splitting down that stalk from the top to close to the bottom, and that would drop a lot of them off on the ground. And, of course, we always had the old saying passed around, 'We're not raising this tobacco for the stalks, we're raising it for the leaves.'"

These days the cutters go down the rows and cut the plants off at the base with a knife, sometimes called a tomahawk, that

Cutting, Housing, and Curing 121

looks much different from the shank knife. The knife, made to be swung like a hatchet, has a wooden or metal handle about twenty inches long. A tobacco cutter works two rows at once. This is termed a stick row. The plants are speared; that is, a steel cone with a sharpened point is placed on the tobacco stick and the plant stem is impaled on the stick. Spears can be purchased in hardware stores for a few dollars. Cutters may use very old, sometimes heavily worn spears.

The conical spear point was often constructed by the farmer. Construction of a spear is described by Paul Carraco. "These spears were made out of a flat piece of steel and cut in a V-shape and then curled around so that it'd have a mouth opening of about an inch, or maybe one inch and one-eighth. And after you'd heated and worked this all down and got it real slick and smooth and had a piece of steel there for a point on it, some people would put a piece of pickup tooth or something like that in the end of it and sharpen it. In spearing this tobacco, we had just a knife that we cut it off with, called them a tomahawk, and it had a blade about, oh, five inches broad at the bottom and four inches up to the top then, and it too had a handle on it where you'd handle it just like you would a hatchet."

Most farmers mentioned that spearing was faster than splitting. Russell E. Wilson now living in Brown County, Ohio, spoke of another aspect of the shift in cutting practices. "The spear came along more or less as a necessity because of the type of tobacco we had. [It] grew much taller. If you split tobacco with a tobacco knife, you couldn't reach shoulder high and do this. It had to be something down where you could reach it. The tobacco that they raised years ago did not grow tall—six feet and so like we have today. It was down fairly short—three feet and it wasn't too bad to use a knife on it. The spear was [a] sort of necessary innovation [for] the type of tobacco that we started to raise."

The number of stalks on a stick varies depending on size. As Paul Carraco described it, you put "five or six stalks on a stick, according to how big the tobacco was. And how good you felt that day maybe, too, whether you could lift those sticks with six big stalks on it. So they were kind of governed in size by the number

Spearing the tobacco stalk. The photo is from a Works Progress Administration collection and was taken on a Fayette County farm in the late 1930s. (Kentucky Department for Libraries and Archives)

of stalks that you put on them, and that, of course, governed by the size of the stalks that you put on a stick."

The cutter starts by putting a stick in the ground at a slant between the row and more or less in the middle of a group of six plants with his buttocks more or less against the middle plant of the back row. In some cases, the field hands have already set sticks; that is, they plant them into the ground in an upright position near where they are working. The way we learned to cut tobacco involved standing alongside the slanted stick with the spear

Cutting, Housing, and Curing 123

placed on it and first cutting the stalk to the right in the row in front of us.

Paul Carraco described the complex movement associated with this process. "In reaching underneath you'd grab the stalk with your left hand and then take your right hand with the knife in it and lean that stalk over with your left hand and cut it off right there at the top of the ground just under your last leaves." When cutting tobacco, we found that it was important to keep the right foot more or less stationary so as to increase the chance of quickly impaling the stalk on the spear without a lot of looking. As Carraco described it from his more experienced perspective, "you had this spear on top [of] your stick and you jabbed [it] in the ground and then you would take this stalk and hold it in both hands, one at the bottom of the stalk where you had just cut it off, you generally had your tomahawk in that hand too where you'd cut it off, and then take your other hand and get down to about middle ways of the stalk, put it up on top of this spear, and shove down on it to split the stalk enough that this stalk would go on down over the stick." Sometimes the stalk splits to the bottom and has to be impaled again. This is called "splitting out."

We followed in sequence with the stalk directly in front of us and proceeded in counterclockwise fashion around the six stalks to be cut. Because the stick is not moved and the cutter needs to maintain some speed, the last three stalks are cut behind him. One risk is that the cutter may cut the back of his right leg (assuming he is right-handed). If the tobacco is small, more stalks will be added to the stick.

Cutting tobacco is strenuous work that requires speed and coordination. It can be dangerous to miss the target when cutting a stalk of tobacco or when spearing the plant onto a stick. Common injuries include impaling one's hand on the spear and cutting one's leg with the tobacco knife. One of us ended up in an emergency room in Carlisle, Kentucky, with a bad cut on the leg. Farm hands are proud of their skill at this task. In fact, there are tobacco cutting contests sponsored in the area. In the banter during the noon meal, cutters talk about people they know of who cut twelve hundred to sixteen hundred sticks a day. Most folks cannot cut

This tobacco has just been cut. As it wilts, the tobacco will settle and become lank. (Kentucky Department for Libraries and Archives)

nearly that much. Proficiency at cutting tobacco requires an early start. People say that if you did not do it as a child, you will never amount to much as a cutter.

Tobacco cutters are subject to the risk of green tobacco sickness, caused by absorption of nicotine through the skin (Gehlbach et al. 1975). The disease and its symptoms of nausea, vomiting, dizziness, and even prostration are self-limiting and of short duration but can easily recur. Workers who smoke or chew tobacco are apparently less susceptible. Because nicotine absorption is more likely with wet tobacco (Boylan 1993), waterproof gloves and clothing afford protection from the condition (Weizenecker and Deal 1970).

Cutting and housing, the most labor-intensive aspects of tobacco production, are done from mid August through September, often the hottest period of the year. Willard Varner stated, "I try to get done by the tenth, fifteenth of September. After the sun crosses the line on the twenty-first of September that's, lots of times, not very good weather. It gets cold, rainy. You can't house tobacco in cold wind. It turns green. You got to get it housed to have heat to

Cutting, Housing, and Curing 125

cure it up. If it gets too late and you don't have heat, it don't cure very good."

Some farmers cut smaller tobacco first because if this tobacco were to be rained upon it "would go down to nothing." By cutting it first, it is also housed first and is stuck in the rafters out of sight of neighbors who might tease about its appearance.

The tobacco sticks are turned over and placed in the ground at a slant so that the cut stalk faces upward with the youngest, most succulent leaves, or the "tails" of the plants, turned upside down and away from the sun to prevent burning. This keeps the moisture in the plant. Then the worker sets another stick and moves on.

The full tobacco sticks are left in the sun to wilt. Wilting is necessary to prevent leaf loss (the leaves break off if the plant is handled while fresh). Wilting also reduces the weight of the tobacco and makes it easier to handle. After wilting, some farmers allow the tobacco to cool down after sundown and put it on the wagon and house it first thing in the morning before the "sun has burned off the dew." If it rains, the tobacco may get muddy and lose value.

These practices have changed through the years. Lawrence Simpson of Fayette County recounted, "When I was young, the tobacco was cut and put in the barn the same day it was cut. The University of Kentucky was promoting leaving the tobacco out for several days. I recall the first crop that we [did] that with. We had cut greener than we would now and it had to be out there about three days and it was brown and all. We went out early in the morning and the dew was piled on it. We had two wagons, teams of mules, and we hauled tobacco all day long. And [with] the last load of tobacco we pulled in the barn it came about an inch or two of thunderstorm rain which would have damaged it terrible if it had been out there. The other people who farmed with me, they thought that was an awful funny thing. Everybody does it now. It worked just great. It was sunburned at a minimum, of course. Tobacco was a lot easier to handle having lost a lot of water."

Good quality green tobacco is quite heavy, therefore, wilting it makes it much easier to handle. Carraco described the benefits of wilting. "The weight of these sticks would be thirty, forty, maybe fifty pounds, some of them, when they're green. If we can get them cut and stay in the field three or four days, then we're driving a whole lot of moisture out and [we] don't have to handle the moisture to get everybody worn out or handling these first heavy crops. You want to kind of get it as light as you can. Of course, it'll go in the barn better, too, and it hangs in there better. If you hang it in there when it's not wilted down good, then you tend to shove it back against the other tobacco [because you have less room]. And if you have a little foggy and humid weather, then you [will] have some houseburning taking place in your barn. That messes up things as far as weight, color, and quality is concerned."

When asked how long the tobacco should be left in the field, Roy Greene replied, "I prefer to take it out to the barn as soon as permissible, without lots of leaf loss. I think if you get a real heavy wilt, the leaves stick together and it somewhat adversely affects your quality. Sometimes in a dry season, you can cut it in the morning and take it in the afternoon. I am very concerned about rain. I think that's a bad result when it rains and gets muddy and the tobacco doesn't have a sheen. If you leave the tobacco out too long it'll sunburn and that's not desirable at all."

The sticks of tobacco are brought into the barn on either a single- or double-axle wagon specially built for housing tobacco with a bulkhead on one end canted at an angle upon which the tobacco will rest. One man drives the tractor, pulling the open wagon slowly through the field of wilted tobacco. One or more men or women walk along the row and hand sticks of tobacco up to men or women on the wagon. The sticks are piled in an interlocking pattern from back to front, which makes efficient use of the space on the wagon and helps to hold the tobacco on. The term for this is to "coop the tobacco." Other workers simply "lay it down." When the wagon is full, the tobacco is taken to the barn to be housed.

An alternate method of wilting and transporting the tobacco to the barn is to load the full sticks directly onto "rail wagons." These have a capacity of up to 250 sticks of tobacco. The plants hang down as they do in the barn. The rail wagons can be left in

Loading sticks of newly cut tobacco, Fayette County. (J. Winston Coleman Kentuckiana Collection, Transylvania University Library)

the field or brought close to the barn for wilting. If the weather changes, they can be quickly taken into the barn. Another advantage of rail wagons is that there is less leaf loss because the plants are handled less.

Various harvesting machines have been developed but are not widely used. Narrators said the machines could cut approximately one acre per day, which is comparable to what can be done by hand. They said it's not that much faster, but it's a lot easier. It's extremely hard work to cut tobacco. Still, the machines do not completely mechanize the process. No one has invented an effective way to spear the stalks onto a tobacco stick with a machine. With one harvester, the operator steers the machine onto a row of tobacco. Then the harvester steers itself down the row and cuts the plants with a saw. The seated operator grasps the plants as they are cut, spears them onto a stick, and stands the full stick onto the row. Some farmers feel the tobacco companies oppose the adoption of mechanized harvesters in burley tobacco because burley is a fragile commodity, so crops harvested mechanically have no flavor or substance, and the leaves are dingy and lifeless.

These harvesters have been in development for some time. George Duncan of the Department of Agricultural Engineering at the University of Kentucky discussed the development of harvest technology during an interview in 1991. "Much work has been done on harvesting machines over the last twenty, twenty-five years. Right now two or three machines are in an experimental stage of development here. One machine is nearing commercialization because a company has decided that they will undertake producing one of the machines that's been developed, which means that farmers will maybe have another opportunity to purchase a piece of equipment for the harvesting of burley tobacco. And I say 'another piece' because in years past there have been three or four different machines that have been devoted to harvesting that have been available for sale. But they have not caught on, they have not fulfilled the needs of the farmers that well, or they were too costly, or some other mechanical problems have all been factors that have limited the commercial success of some of the machines that have been developed here or by private inventors or other companies in the last twenty, twenty-five years. There probably have been eight or ten different prototypes that have been demonstrated. Maybe half of those had some serious and some good engineering behind them."

The barns used for curing burley are distinctive, as are the barns used for curing each of the different U.S. Department of Agriculture types of tobacco (Hart and Mather 1961, 274). Raitz indicates that the origins of this barn form are not clear but speculates, "present barn design is likely the product of adopting the simplicity of the old English side-entry, three-bay barn with the practical needs of proper internal tobacco placement during the cure, access to the interior by tobacco wagons, and a ventilator system that would control the rate of cure" (1991, 18).

The tobacco barn is the central facility used in the post-harvest processing of the crop. Burley tobacco is air-cured, so the barns must be designed to allow plenty of room for air to circulate around the crop. This means the barns are relatively large and tall, with good ventilation and an inner structure that permits the curing plants to hang loosely without wasting space.

Cutting, Housing, and Curing

A tobacco barn, probably from Fayette County in the 1930s. The low building attached to the barn is probably a stripping room. The vertical shutters and ridge-top ventilators are important for managing humidity levels in the barn. (University of Kentucky, Photo Archives)

A sturdy foundation and framework are required to support a large barn and its crop of green tobacco. A Kentucky tobacco barn is supported by fieldstone or cement foundations at the four corners and a series of seven-inch-by-seven-inch wooden posts called barn posts with stone or cement pillars as foundations. Barn posts are placed every twelve feet, the length of the barn. Cross ties that run the length of the barn are firmly attached to the barn posts using wooden blocks. Wooden boxing covers the outer framework and is further supported by knee braces placed on each twelve-foot section. Older barns were constructed using a mix of available woods, including oak, pine, and poplar.

The outside may be painted or unfinished. In central Kentucky, black is the most common color. It is a tar mixture that is both less expensive and more durable than regular paint. Arthur D. Jones described his involvement tarring people's barns. "I bought me a three-quarter ton truck and went around over the country tarring barns and fences. That's what I done mostly all over the county. And out of the county too. Its hard work, but it paid pretty good. I had this creosote, mostly what I used. I'd have to put on two shirts, two pair of pants, grease my face with Vaseline. That stuff would burn you. You had long brushes that you would use. And you'd get to burning. It was rough. That's how I made my living. Hot in the summer time. And especially if you got outside in the hot of the sun and were facing it, it might even burn you up. I done that for quite a few years. When I was working on that tarring, I had my sons working with me. Two or three of them, sometimes a nephew. Sometimes I hired some outside help."

Hinged doors and shutters are built into the outer walls. Tall, thin shutters found along the length of the barn are opened and closed to control air circulation during curing. They also provide light in the barn. Doors, located on the width of the barn, also provide light and ventilation. Each door is supported by a batten, or support brace.

The roof, supported by wooden rafters set on two-inch-by-four-inch or two-inch-by-six-inch plates that run the length of the barn, is constructed of unmilled wooden sheeting. The sheeting used to be covered with tin roofing, but asphalt roofing is now used. Ventilators were often built into the roof, because plans approved by the University of Kentucky Cooperative Extension Service from the 1940s and 1950s included ventilators. This recommendation was later dropped. The floor of the barn is earthen.

The barn is spatially divided by width, length and height. The width of the barn is divided into the "center" and the "sheds." The center is wide enough to drive machinery through. At either end are "center doors" which are large enough to allow the machinery to pass. When both are opened, air can pass through the barn, aiding the curing process.

Sheds extend the length of the barn and smaller "shed doors" are found at either end. Because the roof slopes down from the center, the sheds are less spacious. Ground space in the sheds may be used for storage of farm tools and equipment, and there may also be stalls for animals and loading chutes built there. Most barns have stripping rooms built on as additions.

Tobacco barns are measured in "bents" and "tiers." One may speak of an eleven-bent barn with five tiers." This would be a very large barn and probably an old one, as newer barns are usually lower than this. The term "bent" refers to the space between the

Cutting, Housing, and Curing 131

internal structural support units consisting of posts, beams, and rafters. These twelve- to fourteen-foot areas between each set of barn posts extend the width of the barn. In traditional barn construction, these were the units that were raised in place after being constructed on the ground. Longer barns have more bents. The height of the barn is divided into levels called tiers, which include the entire volume of the width and length of the barn between one tier and another. A taller barn has more tiers. A burley tobacco barn usually has three to five tiers. The top tiers are smaller in volume because of the slope of the roof. The highest area is referred to as "rafters," as in, "He is hanging tobacco in the rafters."

Each tier is marked by cross ties running the width of the barn. Thus, a three-tier barn has three cross ties attached to each barn post with wooden blocks. Within each tier, the cross ties are twelve feet apart. This forms a series of rows across which are loosely laid "tier rails" or "tier poles" for the length of the barn. These can be stripped logs, rounded and small in diameter, or, more typically, milled lumber. The interlocking posts and ties and rails form a lattice framework from which the tobacco plants are hung during the curing process.

The tier rails are unattached to the ties so that they can be adjusted to give the tobacco more or less air space, as necessary. During housing, the tier rails are set approximately four feet apart and twelve to twenty tobacco sticks are hung across two rails. The three-dimensional, multilevel arrangement of tobacco sticks wastes little space yet allows the plants to hang upside down while the air circulates freely among them. If the plants are spaced too closely, the quality of the curing is affected and it is more likely to have houseburn, rotting, and fungus problems during humid weather.

Many very old barns are still in use. The oldest barn observed that was used by a person interviewed retained many primitive features such as a square floor plan, log-pen construction[1] and very narrow centers for the rails. Older farmers reported that log-pen barns were used for tobacco housing earlier. Barns of this type had a square structure of split logs at its core. On either side, a lean-to was added to increase space for curing.

Many of the people we spoke with housed their tobacco in barns that were eighty to one hundred years old. They told stories of barn raising at the turn of the century. The foundations and barn posts were set up. Then the outer walls were built on the ground and raised by a number of men. This was called "pulling them up."

Earlier barns made use of mortise and tenon construction, requiring the higher level skills of the joiner. Spurgeon Louderback of Robertson County described the construction of barns when he worked with L.D. Louderback building barns from 1921 to 1928. Louderback spoke of the strong mortise barns as compared to modern "spike" or "bolted" barns. He recounted a story of a mortise barn on Bee Lick in Robertson County that was blown off its foundation and remained intact. In mortise barns, the major structural parts were fastened together with pins of locust wood. The pins were split from a square block of green wood with a hatchet or froe and mallet. The seven- to eight-inch pins were sharpened and set with a pin driver, which was a metal plate about one-half inch by three by six. The pin driver was set on the side of the post and the pin was driven with a mallet.

After the farmer selected the site, the construction crew leveled it with a pan scraper and a team of horses or mules. The best footers were of poured concrete, although built-up rock pillars were also used. The process of laying out the barn was crucial and required a high level of skill. As Spurgeon Louderback stressed, the person who laid out the barn had to "know the square." After the foundation was finished, the frame of posts and cross ties was laid out on the ground. Referred to as "benting it together," this was anticipated by the cutting of the mortise and tenons with various tools including foot adzes, boring machine, and a two-man crosscut saw. The bents were raised using a crab winch, horses, or tractor. Louderback regarded this as dangerous, or, as he said this is "where the fun comes in." The plates were attached to form the basic rectangular structure. Boxing up, or siding, the barn was done before the rafters were installed. In some cases, the siding was actually cut by the crew on the job. L.D. Louderback had a Nichols and Shepard fifty-four-inch steam-driven saw for this.

Cutting, Housing, and Curing

The rafters were placed on twenty-four-inch centers and toed together. The last part of the process was to install the doors, roof, and ventilators. Mostly the barns had metal roofs consisting of galvanized sheeting. Earlier, "standing seam roofing" was used and featured a seam consisting of L-shaped edges of the sheet mounted against each other and then covered with a cap that was crimped during the installation process. This was replaced with the improved five-V sheeting that involved the sheets being overlapped. The Louderback crew installed "dog-house ventilators" that looked like a series of small houses on the ridge of the barn. Mortise barns ceased to be built because, according to Louderback, "old boys died who knowed how to do it." It is likely that they cost too much.

During housing, the barn hands fill the barn with tobacco from top to bottom, hanging sticks across one bent at a time beginning on the far right or left side. Sticks are handed off from the wagon by the wagon man to a barn man or "hanger" standing on the first tier (or to someone sometimes called a "ground man" or "ground squirrel,") who takes it over to him. He hands off to the second tier, and the stick moves to the top of the barn in this way. On the barn floor, the ground squirrels pick up fallen tobacco or hand off to barn men in other areas of the barn. This job is often done by women. When the man in the rafters has filled his tier rail, the sticks stop at the next lowest tier until it is filled. This continues until the first rails are full from top to bottom. In the meantime, men called "spreaders," who are in the top of the barn, arrange the tobacco carefully on the sticks to give optimal ventilation. In the past, the men would also shake each stick before hanging it. As the rails are filled, it is impossible to climb up to reach plants again until they are "put down" for stripping. Men move over to the next tier rail and begin the process again, filling rail by rail across the width of the barn. When they reach the other side of the barn, they climb forward and work their way back in the opposite direction. They fill the barn in this manner until the entire length is full.

Heavy green tobacco requires special care in handling. The person handling it cannot hold the stick at one end and lift as the

Handing off a wagon, Woodford County. The workers are passing sticks of tobacco up to a third tier. Note how the tobacco on the truck is interleaved. The cross ties and rails for the lowest center tier have been removed until the other tiers are filled. (University of Kentucky, Photo Archives)

stick might break. There is always some stick breakage in housing a crop. When this happens and there is enough left of the stick, the worker may simply leave the tobacco on it and brace the remainder with a "crutch" (usually a piece of broken tobacco stick) placed between the adjacent stick and the rail.

When the barn is full, the tobacco in the sheds hangs down from the lowest tier to within three or four feet of the floor. The center may be left open high enough to allow free passage of machinery. When the entire barn is full, the center is filled so that the plants hang down within five to seven feet of the floor. A few last sticks may be hung from rails stuck into blocks on the barn posts. If a farmer does not have enough barn space for his crop, he may arrange to rent some from a nearby farm. Rental is typically charged on a per rail basis. Like any other resource, barn space can become scarce. A typical situation in which barn resources become scarce and costly occurs in the crop year following a bad crop. As described earlier farmers can carry forward their shortfalls into the next crop year. When most farmers do this, it means they may be producing more tobacco than they can cure. When there is high competition, farmers often are also forced to lease marketing quota in order to get the barn space.

A few farmers practice what is called double-barning. Buddy Rankin of Bourbon County mentioned, "This year we will be even filling some barns twice. We will take some tobacco out and strip it in November or October, and put our last setting in there. This year we will be doing something like that." Members of his family raised more than 150 acres of tobacco.

Climbing in the barn to house tobacco is difficult and dangerous. Each laborer must be able to keep his balance while straddling loose tier rails or poles and reaching down to pull up the heavy tobacco sticks. Occasionally a man "falls out of the barn." That is, he falls from the tier rails to the floor and can be severely injured or killed. Nell Collins talked about the danger of housing tobacco. "The tobacco barns are several tiers high. Some people can't climb up and stand with one foot on one tier rail and one on the other and get a stick of tobacco and pull up. So it has to be capable people because once in awhile a man will fall out of a barn and get badly hurt when they're housing tobacco because it is a dangerous thing, standing up on those tier rails. One might slip over a little, you know, or you might lose your balance. So that's the most dangerous part about tobacco—hanging it in the barns." Perhaps the most difficult job physically belongs to the first-tier man because he has to handle the whole crop stick by stick, lifting it from around his feet to over his head while standing on rails that are four feet apart and not firmly attached. If women are working in the crew, they tend to hand off the wagon.

The trend in newer barns is to build them lower in height. This reduces the size of the crew needed to fill the barn. Retired Extension agent James D. Wells said, "We have local builders that still build old-fashioned four- and five-tier high barns. The problem comes down to labor. We just don't have the labor today that we had at one time. If [we have] it, we can't afford it. So with two people in these two- or three-tier barns [you] can just about hang those barns. Whereas those four- or five-tier barns, you gotta have three or four people up in the barn and one on the wagon. So it just doesn't make sense to build this type of barn." Restricting the height is also a safety measure as it helps to prevent serious injury if a man falls.

As mentioned earlier, the Extension Service advocates plans for nonconventional barn designs. These make use of various innovations including portable frames loaded with tobacco in the field that can be raised into the barn with a hoist system. These innovations are not well suited to traditional barns and the cost of conversion is relatively high (one thousand to eighteen hundred dollars per acre). Also being developed are systems in which the tobacco is hung on a portable frame that is covered with plastic film and left in the field for curing. Some of the systems under development make use of stalk notching rather than spearing. With notching the stalk is hung on a wire.

Once housed, the tobacco must hang in the barn to cure. The plants hang upside down in the open air. The barns are constructed with thin, vertical vents in the side wall that run from the eaves to the ground. When these are opened, they allow the air to circulate

Cutting, Housing, and Curing 137

throughout the barn. The farmer often opens these shutters in the morning during curing and closes them at night to avoid moisture from evening dew.

Curing is a process of gradual drying under controlled conditions done to produce tobacco of the right quality.[2] Burley farmers control the humidity and air supply so that the chemistry of the tobacco leaf changes in the desired way. One relevant factor in this process is the condition of the stalk. The moisture and nutrients in the stalk are slowly drawn into the leaves, where they are used. The stalk helps the curing process.

During the curing process, the leaf's moisture content is reduced from the 85 percent or 90 percent typical of newly harvested leaf to about 10 percent to 25 percent. Newly housed tobacco usually has a green and yellow color. In curing, the green chlorophyll disappears, revealing the color of yellow pigments. Following this, the tobacco turns various shades of brown. This process is more complex than the changes in water content and color imply. Curing is something that happens to living cells. Curing is a biological process in which plant respiration continues and the leaf's food resources are used. It is a "process of starvation" that finally kills the cells. With the death of the cells, the leaf can no longer control oxidation of constituent chemicals, resulting in the final color change to the rich browns of cured tobacco. If the cells die because of too-rapid drying, curing does not occur properly. In these cases the typical green color of the growing plant may be fixed, thus ruining the tobacco. The process also involves substantial weight loss.

During the curing process, tobacco can be damaged by houseburn. This is a condition in which high moisture causes a deterioration of the leaf surface. In mild cases, the leaf becomes "dead" and lacks "body." The tobacco is lighter in weight but the quality is not damaged. In severe cases, mold and bacteria form on the leaf, causing a characteristic odor and discoloration that significantly reduce the value of a crop. Paul Carraco spoke about the houseburn problem. "During the 1960s we had a lot of tobacco that was growing, and [farmers'] barns were built for so much tobacco, and a lot of the people would still try to get this larger tobacco that we were used to growing into the same space in the barn, and this caused troubles. Caused too much moisture in it, get hot, what we call houseburned. [This] is a burning of the leaf by a mold and excessive moisture in your green plants that you'd put in a barn [too] close together. And then they would spread out and just get a real low quality of tobacco and burn so much of your weight out of it." Barns on bottom land are thought to have a much greater problem with houseburn because of their consistently higher humidity than ridge-top barns. Because of these conditions, barns should be built on ridges, with their broad sides across the prevailing wind for the best possible cross-ventilation to aid curing (Hart and Mather 1961, 281, Raitz 1991, 17). Houseburned tobacco was less frequent when farmers split the stalk rather than spearing it. James O'Rourke said, "The stalk remained green [with spearing], holding all that sap clear up until stripping time."

If there is very low humidity, the curing process and resulting color changes are retarded, resulting in green, greenish, mottled (or "piebald") coloration. If the leaf is too dry it dies before the nutrients and other substances are used.

In the past, some farmers burned coke on the floors of their barns during rainy spells in order to dry out the air and aid the curing process. Coke burns without smoke which could damage the flavor of the leaf, and it was cheaper than other fuels. Ira Massie recollected that his father used coke because, "it was so much cheaper than anything else. Back in those days, you could get it for ten, twelve, fifteen dollars a ton, and sometimes you'd get it even less than that if you'd take your truck and haul it."

The heat causes convection currents that move moist air out, replacing it with cooler, dryer air drawn in from outside. Fans or heat sources may also be used to circulate the air. Gas heaters are preferred because they do not produce smoke. Excess use of heat or coke in the early stages of curing can kill the leaf too quickly, causing conditions known as "piebald" and "over coked." Increased energy costs in the 1970s substantially reduced the practice of coking, and many regard coking as a thing of the past. Reynolds Bell described the process. "We were using fires to promote the curing of tobacco. The market did not want houseburned tobacco.

Cutting, Housing, and Curing 139

A central Kentucky tobacco man and his year's work. (Lexington *Herald-Leader*, Staff Photo)

Humid weather promotes a rotting of the leaves, which certainly deteriorates the quality and cuts down in the weight tremendously. The buying trade did not like the houseburned tobacco." Lawrence Simpson said, "There's no tobacco cured with coke in the Bluegrass now. I think it's because they cure it so much more in the field and don't need to. Coke isn't even available commercially any more."

Excess moisture in the air may also cause a condition called "strut." The moist air causes the leaves to protrude from the stalk as though they were walking with a strut. Some farmers burn a powder called "antistrut" to help to wilt and cure the tobacco as well as to retard fungal growth. Silas Cleaver related using antistrut compounds. "You would put like two teaspoons per bent in.

You'd put one in the shed bent and one in the center bent and one in another shed bent if your barn was constructed that away. And so you would have three little spots of fire per bent of barn. This powder would burn, and you could light it. It didn't flame much, just a small glow of a flame, sort of like burning a candle. By the time you got them all lit you were ready to get out of the barn because it would sort of choke you up. Like I say, that was to keep tobacco from strutting. I'm not so sure you can still buy that. I don't know anybody that uses it. [Put] on the dirt floor if your barn was clean. If not they'd take tin cans, just turn it upside down. It didn't burn very long; it soon went out. Maybe ten or fifteen minutes and it was all done. I believe you did that twice a day for some days."

Once the tobacco is housed, farmers turn to other work. Carraco characterized this period. "Whenever it begins to get pliable and the curing up is all done, and then you can go out and do a few more jobs around to get things set up for the next round of tobacco, which will be, maybe, two months or maybe three months stripping. So all of these things have to be figured in as you go along. And the weather will stop you a lot of times. A lot of times you will be thinking about being out in the corn field and working with the corn and maybe soybeans, getting them ready to go up, too. Get them ready to combine. So you have a lot of these jobs that you can do. We have corn to pick and soybeans to combine and then our tobacco to take care of. And, of course, your fall seeding to take care of, too. And, of course, there's generally a little hay around. The second crop or third crop or whatever it happens to be of hay that you want to get in too. And you don't get much time there because you want to get your tobacco stripped and out in as nice of weather as possible."

Many tasks need to be done at this time. Paul Carraco described some of this other work that needs to be done. "Always have to clean up the stripping room too. Your lights that you use will get dusty and dirty and you'll have to wash them down. Then you have to clean out your steam boiler. Maybe you had a little hole rusted in that sometime during the summer and you'll have to get that welded up then before you can go ahead and put

Cutting, Housing, and Curing

steam in your boiler. And then there's your compressor that you have to use in your hydraulic[3] presses. You want to get that oiled up and ready to go. And, of course, you can do that on some rainy days between the time it's curing and the time that you're ready to go to stripping."

In the early fall, a "cover crop" is sown on the tobacco fields. That is, the crop land is "sown down" in wheat or rye immediately after harvest to protect the soil. The fields are harrowed in order to level the stubble and loosen the roots. Winter wheat is usually sown now because rye seed has become expensive and hard to find. Winter wheat sprouts in the fall, "dies back" during the winter, then "comes on again" in the spring. If tobacco follows tobacco on the same field, the cover crop is turned under to reincorporate the nutrients into the soil. If there are not other crops to be planted for the season, the farmer may allow the wheat to mature. The wheat is cut in July. In the past, the wheat was threshed by combine and used to reseed crop land the following year. In recent years, it is more common to plow the wheat under and purchase new seed for the cover crop.

9
The Stripping Room

After the tobacco, including the stems, is well cured, the sticks are taken down. Tobacco is then prepared for the market by stripping the leaves off the stalk, dividing them into grades, and baling them. In the days before highly mechanized corn harvesting, there was often a delay before starting stripping while the corn was cut and put in shocks. Corn cutting, done with a machete-like knife, was time consuming. Stripping is indoor work, so one could afford to wait until after the weather got bad.

This process occurs in the "stripping room." Although stripping rooms vary, most have a long table that stands about waist high. There must be proper lighting, at least over the table. Stripping rooms typically have a row of windows that allow stripping tobacco in daylight or fluorescent bulbs. This relates to the need to clearly see color differences for grading purposes. Hart and Mather note that, "The stripping shed normally extends in an east-west direction so that the stripper is facing north and examining the tobacco by indirect daylight" (1961, 281). This orientation was necessary before electric lights became readily available.

Racks to hold empty tobacco sticks may be set in the center of the room just behind the crew, who work facing the table. Balers are set up along a free wall. There may also be some chairs and, perhaps, a radio. There will most likely be a heater, usually a wood-burning stove because stripping is done from the late fall through the coldest part of the winter. Roy Greene's stripping room, an addition built onto the front of the right shed of his barn,

The Stripping Room

is constructed of wood and concrete blocks and has an earthen floor. A garage door opens from the front to allow easy access.

Dry tobacco is extremely brittle and will shatter if handled. Cured tobacco can only be handled if it is in case. Nell Collins explained, "You have to hang (the tobacco) in the barn to cure, turn brown, you know, and get all the sap out of it, and then there comes a moist time and it comes in what they call 'in case.' That's when the leaf is soft and you can handle it without crushing it." Burl Donahue said, "You can't just put down dry tobacco. You have to have a damp season for it to come in case for you to [be able to] 'put it down.'" A "season" of high ambient humidity and moderate temperature is necessary for the tobacco to come in case. The ambient weather conditions are the key to getting tobacco ready to strip. As Paul Carraco discussed, "Case is something that comes in tobacco [with] moisture. It comes on a warm rainy time. Or, we have a lot of fogs here along the river, and a lot of time our tobacco in the barn will come in case on these fogs." The moisture content of soil and barn, wind velocity, and the amount of "gum" in the tobacco also influence the process. The leaves become soft as the moisture is absorbed. This allows it to be handled without loss of weight through breakage.

    These conditions can vary from year to year. Some years you have to be very patient. Paul Carraco described the process further. "When tobacco is hanging, sometimes the weather will get kind of dry and you'll want to get it in case along October, November, and it'll be dry and cold. You can't get it in case then, so you wait for a little bit warmer weather to come along. And then some of these big fogs that we have, especially if it's rainy season, you have the fog pretty near every night in the fall of the year especially. This has been a pretty dry season thus far this year, so we don't have too much to do as far as getting it ready was concerned. But this past fall it was rather wet, and so this made case getting easy. Pretty near any day that you wanted to, you could strip tobacco out of the barn. In the fall of the year you have to watch [for too much moisture] because there's so much sap in the stalks. If you leave it down more than three days and the big old stalks are green and sappy, then [moisture will] get into your leaves and start it to deteriorating. If you don't watch out, you'll have a bunch of hot tobacco on the floor and then you'll have a hard time getting that in shape [to sell]."

The terms "high order" or "high case" are used to describe cured tobacco that has absorbed too much moisture. Roy Greene explained. "If the tobacco gets too much case, you say it's in 'high case.' It is in an unsafe keeping condition then." A farmer would not put down tobacco that was in high case. If the tobacco was already stripped and packaged in hands or bales, it could easily rot. As Roy Greene said, "The only thing you can do is just scatter it then. You are already in trouble so you try to salvage what you can. You take it out and scatter it, let it dry."

After the tobacco is taken down from the rails, it is put in a bulk and then stripped. Farmers may refer to this as "bulking it down." A bulk is an orderly and tightly packed pile of tobacco stalks with or without sticks located on the barn floor. It must be packed closely to maintain the moisture in the leaves as long as possible; otherwise, they will have to wait for another damp season before continuing with the stripping of the leaves and packaging for the market. Nell Collins explained, "You take it down and put it in a bulk, one stick on top of the other or several sticks—a great big bulk maybe. It'll stay in case longer that way. It doesn't stay in case very long if it's left hanging on the rail. Everybody's always anxious for the tobacco to come in case so they can put it down in bulk and then strip it."

One can pack too much in bulk, however, because it can rot from the moisture and heat inside if left too long. The bulk consists of parallel rows called books. Although it is not recommended, the bulked tobacco may be sprayed with water to help bring it into case. Hot water is thought to be taken up more quickly. Farmers must be careful to not over-spray and damage the tobacco. During this part of the process, some farmers spread straw on the floor of the barn to protect the tobacco from dirt and to help absorb moisture to prevent houseburn.

Stripping includes three processes: removing the leaf from the

The Stripping Room

A stripping room in Fayette County in the 1940s. The tied hands are accumulated on a tobacco stick prior to pressing and bulking them down. Windows provide for natural lighting of the workbench to assist in sorting by color. A tobacco press and pressure sprayer for water can also be seen at the far end of the room. (J. Winston Coleman Kentuckiana Collection, Transylvania University Library)

stalk, sorting it by type or grade, and packaging it. All of these processes have changed over the years, some substantially. Perhaps the least changed is how the farmer actually removes the leaf from the stalk. This has almost always been done by hand, although mechanical stripping machines have been developed. The entire plant can be inserted into the machine, which strips all the leaves off into one grade. Alternatively, one man may strip off the first grade by hand and a second man strip the rest with the machine, or a three-man crew may strip two grades by hand and the tips by machine (Atkinson et al. 1981). A good crew can strip one grade as fast as a machine can. Some farmers using these machines found that they took slivers of the stalk off along with the leaves, giving a messy, undesirable appearance to the tobacco.

As the leaf is removed from the stalk, the farmer often sorts the leaf by type. The terminology farmers use for different types of tobacco leaf are flyings, trashes, lugs, leaf, brights, short reds, long reds, reds, and tips. To a degree, these correspond to categories of leaf type used by the tobacco program to determine price supports. The farmer is aware of the different leaf types he has produced and, to an extent, groups the stripped-out tobacco into a limited number of categories. This only starts the classification process. After the tobacco is delivered for sale, it is graded by a government grader into one of 115 government grades. The tobacco buyer then uses the subsequent processor's notation system as another kind of grade. These notations are quite different from the government grading. A government grader at a sales warehouse showed one of us a company tag with the buyer's notation on it: "X44L, that's a particular grade that he wants to put it in. That's his particular code. I have no idea (what it means)." All of these classifications draw upon some shared understandings about the nature of tobacco and its desired characteristics.

Tobacco grades used to relate more closely to the use of the leaf. Jimmy Bridges explained. "The fine grades are used for cigarettes. As you get up the stalk and get into the reds you get either chewing or pipe tobacco. The leaves on the lower part of the plant are the mildest, and that's used for cigarettes." Currently the relationship between use and grade is not clear as all the different grades are used in cigarettes.

Stalk position is an important aspect of grading while stripping. Roy Greene delineated how grading is done during stripping. "When you strip in grades, the leaf position on the stalk determines the grade. Your lower leaves are your oldest leaves on the stalk and they are more granulated than the leaves [that] are up the stalk. Up-stalk leaves are firmer and heavier. Their leaf structure is denser. The more immature the leaf is, the denser it will be, the firmer the cell structure. There are a lot of criteria [that] go into it. How far up the stalk, the more mature grades, leaves, will be on the bottom. They have been on the stalk longer."

Companies buy the amounts of certain grades needed to make their special blends of tobacco. The buyer bids on only those grades that help to fill his quotas, which vary with the needs of the company.

The approach farmers use to strip into grades has changed a great deal. Arthur Harney Jr. recollected his experiences with changing patterns of stripping. "When I was a boy, my dad would strip his tobacco in ten or twelve different grades. And throw out all the green and all the dark. [He] wanted to strip it perfect. They didn't want any damaged leaves or anything that wasn't right in the tobacco." Paul Collins of Mayslick recounted a similar pattern. "Cigarettes come from the bottom leaves, the trash, and the flyings. They're usually very fluffy, light, you know. After you get up on the stalk, you come to the bright leaf. That was a little brighter and heavier. And then you come to the red and the tips that are a little heavier. We stripped it and made four or five or six kinds. Keep the green out and try to keep the houseburn out. You'd make a long trash, a long bright, and a short bright. And, of course, the trash, you tie that up all together. But then on tips you'd usually make two kinds of tips. You'd make a light tan tip and then the darker tip. Each fella [stripping] would pick his own grade. Usually I was at the head of the table, and I would tie the flyings, 'cause flyings break up pretty bad, and throw the trash under the bench. Then [my brother], he'd tie the bright and pass the red on. We usually had some colored fellow to help us take care of the butt end of it down there. After you got so much stripped, in the evening, when you were going to quit, you'd dig out what you threw on the ground and put it up on the bench and tie them." When Roy Greene was first involved in the tobacco business, most farmers used six grades. "There was a flyings, a trash, a lug, bright, long red, and then short red. And they finally started combining trashes. The lighter side of it went to the flyings, and the heavier side went to the lugs."

In the early days, farmers risked having their tobacco not be bought at all if the grading was not done right. As Ira Massie said, the buyers just walked on by. Aversion to this risk produced a meticulous, even compulsive, approach to stripping. Elaborate grading decreased in time as Roy Greene recalled. "I expect they did that in the '30s or '40s, dropped that, and a few years ago they got it down to three grades. I think that the companies would prefer you did it in three grades now." While we will see that cost of stripping was an important part of these changes some farmers were reluctant to relinquish this dimension of tobacco craftmanship. This is the implication of Art Harney Jr.'s view of this change. "When I started farming, it was beginning to change. I would make four or five grades and strip it pretty. It was pretty well sorted out. And then it gradually got down to three grades, and I never could strip my tobacco in less than three grades. I just couldn't make myself do it. Most people got to stripping it in two grades. But a lot of people make one grade and they can still sell it on the markets we have and probably come out ahead."

Farmers who strip into only one or two grades feel the price differences do not justify the extra work needed to strip into three grades. Brad Carmack, a retired grader, discussed this. "Those are time-consuming things and a farmer thinks he knows, and maybe he's right, I'm not going to answer that, but he don't think it is enough difference in the price to justify him in doing that, so that's the reason you're running across mixed grades. They think they will take less for it and strip it all off in one or two grades. They think it will justify them, or they wouldn't be doing it." When asked if this strategy would work, he added, "Well, I don't know. I always like to get top price out of it if it is top tobacco myself. The best you can do anything is bad enough." The result is that mixed grades are seen more often on warehouse floors. When tobacco is in high demand, a mixed grade may sell nearly as high as a top grade. If demand is lower, the farmer may be forced to return to using three grades to ensure a good price for his crop the following season.

Short production has reduced the number of grades used. "There is a tighter market supply, and so manufacturers need just about everything produced to make their products," suggested Milton Shuffett. "They paid almost as much for the lower grade tobaccos as the higher grade tobacco. And it costs to strip tobacco if you value the labor at the going market price at the same price that you would hire labor. It's costing ten or eleven cents a pound to strip tobacco. And you can strip a little quicker into one or two grades than into three or four grades. And with this short market supply over the past decade, manufacturers have bid prices of the top stalk positions to what they pay for better leaf or better lugs."

A stripping crew requires at least one person to strip each grade. A person often specializes in a particular grade. For example, on the Richards' farm in Montgomery County Oscar Richards' grade is the lugs, and Bill Manley always strips the flyings. When we worked alongside Oscar and Bill, they were set up to bale their tobacco into three grades. The sticks of tobacco are taken from the bulk into the stripping room. The plants are removed from the stick and placed at the far ends of the stripping table, where two stripping lines work at once. The first man on the left is Mike Stull, a farm hand who lives on the Collins Farm. He strips off his grade, which is the flyings. These are the leaves found at the bottom of the stalk. The plant is then passed to the right where the next crew member, Oscar, strips off the lugs and passes the plant to the center of the table. Oscar's daughter, Judy, removes the remaining leaves, the reds. Then she places the empty stalk in the stick rack behind her. This process is mirrored on the right side of the table. Bill strips the flyings, Oscar's son, Doug, takes off the lugs, and the reds go once again to Judy at the center of the table.

During the stripping process, one's hands get covered with gum. Stripping rooms usually have a can of lard or Vaseline to rub on one's hands. This gives the hands a pleasant slick feeling. In the old days, "leaf fat" from the mesenteries of butchered hogs might be used.

The next part of the stripping process is packaging the tobacco for sale. The way farmers package tobacco has changed dramatically since the late 1970s. Currently all tobacco is baled; whereas, earlier, virtually all tobacco was tied into hands. Some farmers sheeted their tobacco, but relatively few have adopted this practice. These different practices had important impacts on labor costs both on the farm and at the tobacco auction.

When hand tying, the stripping room process is organized in much the same way as when baling. A grade is stripped off, and the plant is passed to the next person on the crew, as above. The stripped leaves are held in the hand of that crew member until his or her hand is very full. They are squeezed together tightly near the stem of the leaf and wrapped with a "tie leaf." Tie leaves are leaves of the same grade chosen by the stripper for their length, soundness, and appearance. The tie leaf is folded lengthwise and tightly wrapped around the stems of the "hand" of leaves. The end of the tie leaf is tucked in between the loose leaves of the hand to secure it. The tied end is called the head, and the loose end is called the tail of the hand. Some farmers in the past clipped the butt end of the hand to improve its appearance. The tails are split and the hand is placed over a tobacco stick that extends from the table ledge, held securely by insertion into a square hole cut through the sideboard on the front of the stripping table. When twelve to fourteen hands are placed on the stick, it is removed. Each stick weighs from twelve to fifteen pounds, figuring each hand at about one pound. The stick of tobacco is compressed in a "tobacco press" and then hung from a tier rail or, more typically, placed in bulk in the barn to await marketing. In this case, the bulk is a four-sided, cooped pile of tails-in sticks of tied and pressed tobacco that again helps keep the tobacco in case. Hand-tied tobacco was transported to market on the stick.

Because of the change to baling, tobacco presses are a thing of the past. Usually constructed at home from wood and some simple hardware, presses were attached to the wall of the stripping room. Presses were a popular school shop class or 4-H project. The pressing flattens the hands on the stick which reduces the volume of the crop, helps keep the tobacco in case, and makes it easier to load on a truck.

Kennedy-Galbraith Planter Company of Germantown advertised their Gateway tobacco press for stripping-room installation in the mid 1920s. Advertisements of it made the following claims: "It will make your tobacco look so nice that you will be astonished at the change in its appearance; it will make your tobacco bulk down easily and neatly and in much less space; it will add 10 percent to the weight of your crop because good pressing and close bulking holds the case and prevents drying out; it will save

The Stripping Room

A tobacco press. Presses consist of a flat wooden surface fixed to the stripping room wall, a second movable wooden platform hinged to the other at the top, and a lever for bringing the two surfaces together to compress an entire stick of tobacco. There is also a bracket to hold the stick in the proper position for pressing and a counterweight to help raise the movable half of the pressing board. The pressing helps keep the tobacco in case and makes it less bulky for transport to market. (University of Kentucky, College of Agriculture, Communications)

you time and unnecessary rehandling." It is unclear if presses were available before the Kennedy-Galbraith press. Some people made use of larger capacity presses, called wagon-bed presses. It was possible to press fifty or sixty sticks at once in such devices, we were told.

In baling, each grade is placed in a pile on the table as it is stripped. Another crew member places these piles in the appropriate baler and "runs them down" as needed. He or she also brings

A tobacco baler. As baling became universal, specially designed equipment such as this became available. The pneumatic compression device can be seen to the right of the operator. A close look reveals the carefully placed cotton strings used to tie the bale. (University of Kentucky, College of Agriculture, Communications)

in sticks from the barn and replenishes the supply of plants to be stripped. This person empties the stick racks, piling the tobacco sticks in the shed and loading the stalks on the wagon. He or she ties the bales, removes them from the balers, and carries them into the barn, where they are stacked in bulk and covered with plastic until the farmers are ready to transport the tobacco to market.

The bales are rectangular and weigh from fifty to one hundred pounds. Jimmy Bridges described the process. "Now they're going to the baling machines. They're pulling the leaves off and turning around and putting them in these balers, and they have a compressor, and it bales this tobacco, and they tie them up with string, and they take them to market that way, each bale with the same grade; you can't mix it."

The "baler" can be constructed by the farmer at little expense using plans published by the Extension Service. A wooden box is constructed. It is three feet long, one foot wide, and two or more feet deep. Notches are made to feed three strands of twine along the insides of the box. These will be used to tie the bale. A wooden lid is constructed so that it fits inside the top of the box and can be pushed down with a pneumatic cylinder to compress the leaf. Specially manufactured baling equipment is now widely used.

The leaf is placed lengthwise in the appropriate box with the stems pointing toward the outside. The bunches of leaf are placed alternatively pointing to the left or right with the stems abutting the wall of the box. When the pile of leaves gets close to the top of the box, the lid is pressed down on the leaf. More leaves are added, and the pile is again compressed. Some farmers prefer to "run them down twice" in this way to make the bale firm and compact. Loosely packed bales can fall apart during handling. When the bale is large enough "from elbow to knuckle" (twenty-two to twenty-four inches high), the front of the box is taken off and the three strings are tied with bow knots to secure the bale. The string, by regulation, is cotton because manufacturers are concerned about synthetic string burning in subsequent processing.

Baling is now universal. Farmers agree that the main advantage to baling is that it saves labor. "It is faster, and when you save labor, you save money," Roy Greene said. Virginia Calk added, "The main thing is it just takes less help. You can't get farm help and it is so expensive now."

The University of Kentucky Cooperative Extension Service began to promote baling as a labor-saving innovation in the late 1970s. It published plans for the construction of baling machines and held educational meetings to describe the advantages of the process. An experimental marketing program began in the 1978-79 marketing season that allowed a limited amount of a farmer's crop to be eligible for price supports when packaged in bales. The amount that could be baled increased to 25 percent in the 1980-81 season (Atkinson et al. 1981), and by 1982, the whole crop could be baled. Although some innovators began to bale their tobacco right away, most of the farmers waited two or three years to see if the new idea would be of benefit.

Farmers were heavily involved in the development of innovations in packaging tobacco. James C. Rankin of Bourbon County recounted his involvement in the development of baling technology. "Up until we came up [with] baling, loose-leaf tobacco [stripping] never changed from when the Indians gave it to us. You pulled it off and put it in a hand and tied it. We started trying to get that changed in the '70s. I took a vacation to see my brother in South Carolina. It was in July and that was their marketing season. My brother took me out. He said, 'Let me take you to see how they market flue-cured tobacco.' They had it all in sheets, loose leaf. I told them, 'You all got us beat all to pieces on this marketing.' They said, 'What do you mean?' And I knelt down on the floor and showed them how we had to tie our tobacco up in a hand. They took a picture of me showing them how to do it and put it in the local paper. I came back off of vacation, and I called our county agent. I said, 'We have to do something.' He said, 'What are you talking about?' I told him I just came back. 'That will work in burley. What those people are doing down south will work in burley.' He said, 'Lets have a meeting.' So we called a meeting down at the courthouse. And had a big crowd. He just put in the paper 'meeting at the courthouse about burley tobacco.' Ware-housemen come. And he started a meeting, and he turned it over to me [and] told them, 'Bud Rankin wants to try to get something going.' Well they all laughed at me. All but about five or six. The warehousemen wanted to run me out of the country. Five or six of us hung around that night. I told them what I had in mind. If I keep raising this tobacco, we have to do something we can't keep tying this tobacco back like when the Indians gave it to us. Then I thought the sheets was the way, the way I wanted to do it, like they did down south. I stuck with that idea; I was hardheaded. Stayed with it for four, five years. I haul my tobacco to

Tennessee for about five years because there was a warehouseman down there who supported it like I did. I lost a lot of money. I hauled it down there. And took a lot less money for it. So Bernie [the county agent] stayed right in it and contacted the university. And it so happened that the university had this closed deal that they were playing with it. They had give up on the sheets and they were going more to a baling box. It's a box about eighteen inches one way and thirty-six inches the other way, and it's got a top on it. So Bernie contacted George Duncan and Joe Smiley at the university. They said, 'We are already working on something. And we would be glad to work with your people.' So we were the first. I think there was five of us. So right out there in that barn, Joe Smiley brought the first box that was built and put it in that barn right out there. That year, I am not sure, but I think [there were] five different apparatuses. They got Philip Morris Tobacco Company involved in it. We finally sold the idea that if we continued to raise tobacco, we had to do something about [stripping costs]. We could not continue to pull it off leaf by leaf and be productive. Philip Morris, they were great. They put a lot of money in it. They had to donate it to the university and work through the university on it. They guaranteed a certain price for our tobacco. This is after I had hauled tobacco to Tennessee for a long time. And the original first five or six had priority of being in the program, and we picked up some more people who got into it. They had to have a lot more tobacco. And the first program, I think, if I am not mistaken, they would come up with so many thousand pounds of tobacco that we would put into sheets and so many thousand pounds that we would bale. And that a way they could analyze it on the marketing system. I was prejudiced toward the sheets. I was really disappointed that I was losing ground. From my personal [perspective], I would still rather have sheets. However I got to realize that the warehouse has a problem [with] them. The warehouse fought us pretty good on it."

Sheeting involves placing the leaves on a spread-out fabric sheet (eight-foot-by-eight-foot square in size) supplied by the warehouse. When full, an amount weighing approximately 150 pounds, the corners are drawn up and tied together. The attitude toward sheeting seems to be generally negative. It has been described as sloppy and too bulky.

All parties seem to have objections to sheeting. Farmers consider it messy (although some use it for collecting the scrap tobacco found on the floor and table after the day's stripping). Buyers don't like it because it is more difficult to examine the overall quality of the leaves in that it is easy to conceal bad tobacco with this kind of packaging. Companies do not want to have to prepare lines in the aftermarket processing plants for three types of tobacco packaging. Warehousemen dislike it because the tobacco is least compressed by this method of packaging and the sheets cannot be piled on top of one another so they take much more warehouse floor space to display it for auction. Many warehouses have a fee for each lot sold in addition to commissions based on weight.[2] Obviously, it is to the farmer's advantage to put the maximum poundage on each lot. Since only 150 pounds of sheeted tobacco will go in each lot, as opposed to up to 700 pounds of baled or tied tobacco, there is little economic incentive to adopt the practice. While sheets are the least appealing packaging method from the standpoint of aesthetics, they are the easiest to do.

The University of Kentucky College of Agriculture took an important leadership role in these innovations. George Duncan discussed the development of baling technology from his perspective as an agricultural engineer at the university. "An agronomy Extension specialist and [I] helped [get this] idea started in 1974, along with others in the college that helped evaluate some baling technology. A couple of us who were actively involved in new ideas like this said, 'Let's try something. Let's try it in an organized manner. Let's get people to look at some alternatives.' So we began experimentally testing the sheet method of the South. Canada had gone to a bale method in the meantime. It was in the mid 60s that Canadians switched from hand-tied to a small fifty-five-pound bale. Some other tobacco-producing countries of the world had different sizes of compressed bales. So the bale was not new around the world, but it was new to Kentucky people. In '74 we began to methodically work with industry, with some farmers, and did the first experimental work with thirteen farmers of Bourbon County to experimentally package some tobacco in a bale form using a specially built plywood box and a means to compress a bale of about seventy pounds. That met with very good reception with the farmers. The tobacco company that handled that tobacco said, 'Well, it has some possibilities, but we need more work.' So the next year we expanded it, and the next year it even expanded further until we got up to about a three-million-pound experimental project. And that's a whole lot of tobacco, with 235 farmers in about fifteen or twenty counties. We had a big project comparing methods, getting data, evaluating—a lot of people involved. But there were certain political segments of the industry said, 'We don't want to change. This looks like it could be a detriment.' So there was a stalemate then for two or three years until some of the political efforts were resolved. And finally by 1981 some of the efforts had been resolved, the feasibility of packaging tobacco loose leaf had been evaluated, and, in short summary, the bale method was accepted and began to be used in '82 and '83. And in about three years, we had over a 90 percent conversion factor. Probably the most significant change of any one commodity involving hundreds of thousands of people. It took a tremendous amount of educational effort. There were publications and drawings on how to build a bale box. There were videos. There were meetings. Many efforts in that period that really enabled the transition from hand-tied bundle methods of handling tobacco to the bale form. Of course, now [bales are] standard and most people don't realize how that came about. But it was a great revolutionary effort from 1974 until about 1983 with tobacco leadership and industry and companies all jointly working to evaluate and come to an acceptable conclusion as to what ought to be done. Some were reluctant to accept it, others pushed it hard enough that it came about."

Farmers saw big advantages in baling. Oscar Richards described his switch to baling. "Now they had allowed baling for a certain percent of your tobacco for the past three or four years. You could bale 15 percent or fifteen hundred pounds of your crop, and I wouldn't set up for just that little bit. When it got so you could bale it all, I went to baling. It is so much faster, a lot faster.

It's an advantage all the way through. You strip your tobacco drier. It's kind of hard to get tobacco in the right case. You go to putting down and it will be a little dry or it will be higher case or something. Well, if it is dry, you can strip it and put it in this bale where you couldn't strip it and tie it at all. You have to have a tie leaf[1], you see. If the tobacco is dry, you can't find a tie leaf and you got to hold it in your hand, too, to tie it. This way you just snap it off and lay it on the table and put it in the bales. So you can strip tobacco when it looks drier. That was a big relief."

Some farmers objected to the practice. Some thought baled tobacco would bring a lower price at auction. Others said it would be difficult to keep tobacco from rotting in the bales, and some objected for aesthetic reasons. The aesthetic dimension is important. A letter from a tobacco farmer that was published in the Lexington *Herald-Leader* in 1981 stated, "This is the worst display of tobacco for market I have seen in my 38 years of growing burley. Packaging tobacco in bales and sheets takes all the pride out of growing a quality crop of tobacco. I suppose this is the way of the world, and what some people call progress. To me, it's taken all the pleasure out of growing and sending a top quality crop to market" (November 21, 1981 Tinnie G. Carr). Another objection to bales was that they are easy to steal. They often weigh from eighty to one hundred pounds and are therefore worth more than $150 each. In response, the university came up with a press to mark an identification code on tobacco leaves, which can then be scattered throughout each bale.

Tobacco companies were concerned that packaging tobacco in bales would lead to easy concealment of lower grade, poor quality, or green leaves within. To grade a bale, it must be broken open, but usually only the top bale of each lot is examined. A grader told one of the authors, "They untie one (bale) on top. You can glance down inside and maybe see some inferior tobacco down in the bottom and maybe you don't." Farmers who bale their tobacco must now sign a statement promising not to deliberately mix the leaf in their bales. They take this seriously for if found in violation of the agreement, they lose the support price on their tobacco.

During the early 1980s when baling was first being used, stories circulated about farmers putting rocks and other things, such as dead animals, in the bales. It is unlikely that this happened much. One farmer we spoke with said he "saw with his own eye" a bale that when cut open revealed stalks cut into thirds in the bale.

For a time, hand-tied tobacco and baled tobacco brought different prices. Oscar Richards explained. "The first two or three years that they started it, I'd say that there was a nickel difference in the price. Year before that it just fluctuated. You'd go along and one day they'd pay a nickel less for it, say today. And then tomorrow they'd pay the same price for it. And then maybe next week they'd come along and drop it." But by 1980, the tobacco companies had accepted bales and were bidding as high for baled tobacco as for that tied into hands. Virginia Calk described the change. "Nineteen-eighty tobacco was the first that we had ever baled, and there wasn't but maybe a cent or two (per) pound difference in the price. But [in 1981] the buyers paid the same for it. One of my tenants had some baled and some hand-tied on the same row and they brought the same price on the market." In 1981 a farmer was quoted in the Lexington *Herald-Leader* expressing his resentments about this: "We hand tied and graded it in three grades just like the companies said they wanted it. We thought it should have brought a little more money than the average for the work we put into it."

Once the market price was found to be equal, baling was rapidly adopted. By this time, farmers could see that storage was no more of a problem for bales than it was for hands. The problems with rotting tobacco occurred during periods of high humidity in either case. Farmers now like bales because they are labor saving and easier to handle during storage, loading and transport to market.

All of the tobacco leaves are marketed. Even the scraps of shattered, rotten, green, or sunburned leaf that are thrown under the table during stripping are gathered up and packaged together. Some farmers call green tobacco leaf that is broken off the stalk during housing "ground leaf." Ground leaf is almost always dirty. The very light tobacco at the bottom of the stalk that is not tied immediately is carelessly placed under the table and called, in some localities, "dog bed tobacco." The low-quality, mixed-grade price this leaf brings is still better than nothing.

The remainder of the plant is the stem, or stalk. Burl Donahue said he would be willing to sell the stalks, too, if he could. Until there is a market for them, he will use them as green manure on his fields. Every once in a while, one can observe tobacco stalks scattered on lawns in Lexington as fertilizer. During the winter, the stalks also provide traction for the trucks driven on the farm to do chores. Farmers with black shank or other disease problems do not usually recycle the stalks in this way.

10
On the Floor

Burley tobacco is marketed through a loose-leaf auction system that was first used west of the Alleghenies at Clarksville, Tennessee, in 1901, based on a pattern established at Richmond, Virginia, in 1842 (Clark and Browning 1953, 5). It is referred to as a "loose-leaf market" because the tobacco is not packed in containers. From the earliest days of tobacco production in the 17th century, leaf was sold in hogsheads because of the distance tobacco had to be transported. In the early days, most tobacco was shipped to England. After the Civil War, tobacco was marketed by auction in hogsheads in places like Louisville and Cincinnati. The hogsheads were assembled on a sales floor and then opened for inspection. This tobacco was sold "on the breaks" because inspection involved breaking into the hogsheads. Even though tobacco could be inspected, this system was abandoned because buyers could not determine the quality of the tobacco. As Tennant stated, "The perennial difficulty of dishonest packing was present even when the buyer himself attended the inspections" (1971, 211). Loose-leaf sales were also encouraged by "improvements in transportation, the more rapid growth of domestic manufacture, the growing importance of flue-cured tobacco, and increasing concern about tobacco grades" (Tennant 1970, 212).

The Kentucky white burley market opens in late November. The warehouses accept tobacco in early November or late October in some years but cannot officially weigh tobacco until ten days prior to the "opening sale," usually held during the week of

Thanksgiving. Farmers begin taking tobacco to the warehouse at this time and continue until a month or two after the first of the year.

Because tobacco is bought and sold by the pound, farmers are interested in the factors that affect the weight of their crop. There is a difference in the weight of some varieties. "Some types of tobacco weigh heavier, some are lighter weight, and you never know which is going to be the popular one on the market," explained Virginia Calk. Other factors that affect weight are the time of year and the amount of humidity in the air. The less water there is in the leaf, the less it will weigh. Weight loss continues after the tobacco is stripped and packaged for market. As a rule, then, it makes sense to sell the tobacco as soon after stripping as possible. According to Roy Greene, "The longer the tobacco hangs in the barn, the more weight loss. If you strip tobacco in November, it will weigh more than if you stripped it in January or February because there is a process that just eliminates weight. It tends to dry up more." The weather is a factor that may interfere with this maxim. Greene explained, "You do get the advantage of it weighing more [when it is humid]. Of course, you can strip it real dry and handle it real dry and take it into the warehouse in November and probably have some weight loss through shattering and drying. Take a crop that is real high order in January, it would probably weigh pretty good. There are variances there that you can't rule out."

Some people believe you can increase the weight of the crop by taking it to market when it is in high case, or by actually causing it to absorb moisture. Oscar Richards disagreed. "No. Now it would be too little, so many people has got the wrong impression of that. A lot of people ruin their crop of tobacco with a tank of water. They show you there's not much difference.[1] It seems funny now, you can pick up a leaf of tobacco that alaying there that's bone dry, now you can't tell if you got anything in your hand. You can pick up a high case leaf of tobacco and it seems like it is a little [heavier], you can feel the weight, but it's not that much difference."

Some farmers like to get their tobacco to market early because it gets in their way. "Sometimes [tobacco] gets in their way in the

Street scene in Lexington during the 1924 tobacco market. Trucks and wagons are lined up for unloading.

barn, and they just don't feel safe with it laying around in the barn," said Paul Carraco. "This tobacco is moved on the floor a lot of times even before it's officially opened. You can generally get on the floor with tobacco by the 1st of November. It takes up room in your barn and you want your barn [space] to spread more of the tobacco down."

Some farmers haul their tobacco on their own equipment; others hire someone to do this. Arthur D. Jones described hauling tobacco to the Mount Sterling market. "You'd have to get someone to haul it. They had people that trucked around the country. That was their job to haul tobacco. They'd load and you would hand it up to them and they would take it to market. You'd have

Unloading tobacco at a Lexington warehouse in the early 1930s. Two men are making baskets. (University of Kentucky, Photo Archives)

to pay them so much a load for hauling it. I remember a time or two, the landlord I was on, he had a truck and I hauled my own tobacco a few times. You have to put that tarpaulin on it and tie it down good because you know before you get it unloaded sometimes you'd have to wait in Mount Sterling. Now you hardly ever see a load covered up because it don't take but a little while to get where you are going. These bales made it altogether different because [with hand-tied tobacco] they'd have to take it off a stick at a time and put it on baskets, and it took quite a little time to put it on. Now that same amount of tobacco you can put on in a few minutes."

In the past market prices were a major risk factor, but the tobacco program has controlled this to a large degree. Because minimum prices are set, there is not a great risk of loss when the tobacco is sold. However, there are fluctuations in price. Some

On the Floor 165

farmers believe it is best to sell when the market first opens in November. Others don't think it is worth the rush to finish stripping early. They prefer to watch the market for awhile and trust to a high quality product for top prices. Some state that certain towns and warehouses bring better prices, but most seem to sell close to home in a familiar warehouse.

The opening sale of a market is special because of the anticipation of the first sale and the excitement of the determination of prices. Local officials may show up to be part of the event, and no doubt local merchants are hopeful of seeing some of the "tobacco money" later in the day. The warehouse that has the first sale is different each year, based on a rotation system.

Some farmers work as solicitors for warehouses. They try to get their neighbors to sell their tobacco at a specific warehouse. Farmers may be offered various incentives for taking their tobacco to specific warehouses. These include free seed, barn-to-warehouse cartage expenses, free lunches, and rebates on the selling charges. Warehouses also vary in terms of their sales charges.

Until the early 1980s tobacco was taken to the market packaged in hands, of course. This required some special techniques. At the warehouse, hands were removed from the tobacco sticks and arranged in a circular pattern on "baskets" by a "basket man" who sat in the basket and placed hands in a ring around the outside edge. Then he filled in the middle. The tied ends of the hands (heads) are placed around the edges of the basket, with the loose ends (tails) lying toward the center. Roy Greene said the secret to creating a perfect basket of hands is to spread the heads on the basket and to keep the tails close together. Several layers, all of the same grade, are stacked in this way until the arrangement is four to five feet in both height and diameter and weighs no more than seven hundred pounds. The effect is beautiful, especially when the tobacco is neatly tied and of high quality and without mud or spotting.

Making a basket was done in the presence of the farmer, who, within limits set by warehouse policies, decided how the tobacco would be grouped. The basket man stacked tobacco as it was brought to the basket by the producer. Lower quality tobacco was placed toward the bottom of the pile, although the lowest quality was not put on the bottom. The quality of the tobacco toward the top was better. The very best tobacco was placed on top. In unloading, the farmer might reserve two sticks of good-looking tobacco to serve as cappers. Because different grades of tobacco were placed on different baskets, the weight of baskets that comprise a farmer's sale varies substantially.

Paul Carraco described the process. "It would be put on a basket. You'd pack in a round cylinder shape by taking the tobacco and have one person handing [in] tobacco, one man on the basket on his knees. You'd run this in a circle around there and on top of it, pressing it down on his knees all the time to keep it packed down together. You'd go clear around the outer circle, and then you'd have a space left in the center. You'd have to add some hands, just enough to tie that together. If your tobacco was shorter, you could add a whole half a stick at a time in the center of the baskets to hold it together. Short fluffy tobacco was hard to handle [and] didn't stick together. You might think you were doing all right. First thing you know, if somebody bumped the basket a little bit the whole thing would pop open. You get some good old long bright leaf and lugs, they were long enough and generally in case good enough, that you didn't have to worry too much about them. They'd lay right in there and stay in place like they should be."

At the warehouse, tobacco is unloaded and segregated by grades. In the case of hand-tied tobacco, our informants spoke of it being "stick graded." Bales are stacked on a pallet or cardboard "slipsheet." They are strapped onto the cardboard with metal tape for easy transport within the warehouse. Many warehouses used baskets for bales initially, but they were less stable because the bottoms are slightly curved. As many as eight bales may be stacked up, as long as the weight does not exceed seven hundred pounds. Paul Carraco described how the bales are dealt with at the warehouse. "These bales are put off on pallets on the warehouse floor. It's seven to eight bales to each pallet, and maybe the pallets will weigh around six hundred. You can't get them over

A basket of tobacco on the scale. Note the duckbill dolly. (Kentucky Department for Libraries and Archives)

seven hundred pounds. And this is put off of your truck onto the [forklift] and [they] carry it to the rows and put it where they want it."

Unloading baled tobacco is much quicker than unloading sticks of hand-tied tobacco. This has resulted in a reduction in hiring seasonal help in the warehouses. A manager of a Lexington warehouse said, "Tattersalls used to have sixty employees with the hand-tied system. Now they can do the same work with five or six" (Osbourne 1990, D8).

After the tobacco is put on pallets (or in the old days on a basket), it is weighed, recorded in the warehouse records, and placed on the warehouse floor. The farmer is given a receipt that identifies the grower, lot number, and number of pounds. This places the tobacco under the responsibility of the warehouseman in case of fire loss. The pounds are tabulated and recorded on the

Tobacco lined up for sale under the skylights of a Lexington warehouse in the 1920s. Some tobacco is capped with inverted tobacco baskets. Weight/grade tickets can also be seen. (University of Kentucky, Photo Archives)

marketing card when the tobacco is weighed. If a producer has more than his quota, the tobacco is never lined up for sale.

Tobacco is lined up in rows on the warehouse floor with enough space to allow the buyers, farmers, spectators, auctioneers, starters, and ticket markers to pass. The buyers are on one side and the auctioneer is on the other looking across the row to catch their bids. The tobacco is sold rapidly.

The tobacco piled on a basket or pallet is considered to be one lot for the purpose of selling. Each lot is piled up and weighed when the farmer delivers his crop to the warehouse. A warehouse is allowed to sell a certain number of lots each day by the amount of floor space.

Warehouses charge the farmers fees for their services. Each has its own system, but many charge per hundredweight sold and/or a basket charge. This pays for the warehouse facility, supplies, warehouse equipment, labor, advertising, and profit (Shuffett 1986). Warehouse fees in burley are not regulated by the state or government price stabilization program, but they are state-regulated in the flue-cured belt. The warehouse plays a major role in the administration of the tobacco program by keeping track of poundage sold and assessing government taxes on each farmer's sales receipts. Warehouses also play a financial role in the community. We heard many stories about warehouse treasurers helping a tenant to purchase his own land or to make it through a poor season. Regular customers may receive loans from the warehouse during hard times.

Berle Clay of Bourbon County discussed these relationships. "We hauled the tobacco to Paris but also to Winchester and Cynthiana. And I think that depended on two things. You always tried to give the tenant some freedom to sell where he wanted to sell. And depending on where he came from, he might have ties to another warehouse where he had sold tobacco before or his parents had sold tobacco. I know in certain cases my dad would get an agreement with a tenant to take tobacco to another town like Winchester or Cynthiana. This is a very important relationship between the warehouseman and the grower. [The] relationship between the warehouse and the grower is not simply the tobacco relationship. I remember when we combined wheat and clover, we always poured it on a warehouse floor where we sold tobacco. In other words, because we were steady customers at the warehouse, we could use the warehouse other times of the year. It was a very paternalistic relationship between the warehousing establishment and the tobacco grower. Obviously the tobacco crop was a big source of money for the small farmer. It was a very important, critical sale, and typically the whole family would go to watch the sale. Both [the tenant and the landlord would] go. You know, the check was split half and half. He had to pick up his check, and the tenant picked up his half.

"Warehouses compete for who pays the highest price of the day. Generally that highest price is a basket that the warehouse has bought itself to create the impression that it's paying the highest price. They turn around and sell it later on, but they can take a slight loss on it. But the highest price was obviously regarded as a measure of how successful the warehouse is."

Tobacco farmers are proud of the careful handling of their crops, and the display of their season's labor on the warehouse floor is a part of the annual ritual. When walking on the warehouse floor before or after an auction, one hears comments about the tobacco on display. Farmers compliment well-handled crops and criticize careless production technique while they analyze market trends.

Sometimes the condition of the tobacco on the floor may be poor. Paul Carraco described this. "In the warehouse you see a lot of different kinds of tobacco. You see tobacco that has been left outside and it got dirty and rained on it. Mud will wash up on it. You see [tobacco] where it's been put in the barn too tight or wet. You see tobacco where it looks so ornery and black. You just wonder sometimes how you can face tobacco after seeing it like that and wonder if it'll even get in position where it would be in a nice white cigarette paper or in a package of plug tobacco or in a package of scrap tobacco where it was all fixed up and looks so nice and all."

In 1931, after overwhelming endorsement of a referendum of farmers, the Tobacco Inspection Act was passed, creating the current scheme of leaf grading. Official standards for leaf grades, largely based on those of the pool, were developed. From the beginning, leaf type, quality, and color were important.

Before the tobacco auction begins, the government grader goes down the rows of lots to be sold and gives each a grade which describes the type of leaf, its quality and its color. "First thing you establish is the group," said Roy Greene. "That's your X, C, that's in the first cycle. The next is the quality. That's your specification. And third is color." The government grading system judges stalk position, quality, and color separately. The three are combined to give the final grade. Thus, many more combinations are possible. All have a set support price except for three inferior grades.

The process of grading is complex, as indicated by Jimmy Bridges. "Now your Uncle Roy was a tobacco grader on the floor, and he put the grade on it to describe that tobacco. Each pile of tobacco would have a grade on it. He'd put a grade on it whether it was first quality and the kind of leaf it was, like X3F would be flyings, third grade and the F would designate the color. An F would be a tan color. If it's red tobacco, he'd put an R on it. He'd have to go down through the warehouse and put that on every basket. And he'd write on a ticket what it is and what the government pays, they have a floor on tobacco. This year X3F would be maybe $1.80 something. Then the buyer, in order to get that kind of tobacco, would have to bid a little over or else the government would give you that money for it."

Greene explained how to grade tobacco. "The stalk position of the leaf determines what the grade is. The lower leaves on the stalk are the oldest leaves and they mature first. You have your X's and C's and T's and B's. X's are your lowest leaves on the stalk. They are the trashiest, thinner. The higher the leaves grow on the stalk, the denser the leaf structure is. The C's are your lugs. They start in the middle. They are usually the longest and are the widest leaves. B is your leaf and T is your tips. [The leaf] has a lot in common with your lug. It is not quite as wide but it is one of the most desirable right now. It used to be the lug was the most desirable, but I believe dollar-wise a good leaf is more desirable."

The color of a tobacco leaf corresponds to its maturity and stalk position. According to Jimmy Bridges, "It's green when you cut it, and when it cures in the barn it's tan or straw colored—anywhere from straw to red. The straw colored, of course, is the best part. The lower parts on the stalk are straw colored and as the leaves get up to the top they get red and they're sort of a reddish tan in the middle if it's cured properly." Roy Greene explained the criteria government graders use to grade color. "Your flyings and your lugs are your lighter bodied tobacco, and they are also lighter colors. When you got your B's and T's, there are no L's, and F is your lightest color. If it is reddish tan, it is FR, then the darkest is D. Of course, you have your greenish red and that's a V, like VFR, VR . . . I mean it's greenish tinged, around 20 percent of the leaf or less. Now if it is over that, I think you go to the G. If it's greenish, it's VF or VR; if it is green, definite green, then it's GR. In flyings and lugs, in X's and C's, you have L is the lightest color. Tan is next. X3L is just a shade lighter than the normal color. F is tan. Then you get up in your leaf and also there are colors that denote other qualities, like green (G)." And variegated is K. Variegated is undesirable. Green is not desirable. Under present requirements, I believe F is the most desirable color. You don't find many R's or FR's in leaf and lugs, I mean trash and lugs. X's and C's."

The grader next determines the quality of the tobacco. This is indicated by the numbers one through five. Greene explained, "There are five properties in each group. One is your best property through second. Third is really in the middle. Four and five are more inferior and less desired."

The grade determines the price at which bidding begins. If a company buys a lot of tobacco that is then found to be unsatisfactory, the sale may be rejected before the tobacco is taken from the floor. A grader explained, "Now, if there is something wrong with this pile of tobacco when they get it in the co-op, it should be drawn to this grader's attention. His initials are on there, and they can refer back to that and come back and say, 'You made a mistake on this pile of tobacco. It should have been graded NOG or B4R or some other grade.'"

As mentioned earlier, some farmers strip into mixed grades. While this may be done carelessly by an inexperienced farmer, it is more likely an intentional strategy. A mixed grade carries a lower support price but may bring a high bid in times of high demand. If the market is good, a farmer may strip into only one or two grades to save time and labor, hoping that the price received will not be so low as to cancel out savings on labor costs.

With hand-tied tobacco, it was possible to conceal a few leaves of a lower grade or quality in a higher grade hand. Farmers may call this "blending" the tobacco, as one told us: "Well, I pull them off and then I twist my hand, turn it a little bit. I get the next few leaves that would be heavier and darker, and the next stalk I put the light grades right on top of them and then I switch my hand a little and put the heavier grade in the middle. And that way the heavier grade is always in the middle, the fine grade is on the outside. [My friend] used to laugh at me doing that and I would come to a green leaf and he would say, 'Oh, you'd better throw that out.' I'd say, 'Oh, they'll never see that,' and I'd just work it right on in the middle."

Rules of thumb govern the decision-making process when mixing grades. It is generally accepted that "a lower grade will carry a higher grade." That is, one can put leaves of a higher grade to a hand of a lower grade without risking a mixed-grade price. One farmer said, "leaf will carry lugs, but lugs won't carry leaf; C (lugs) will carry X (flyings) but X won't carry C."

The goal of these stratagems is to receive the higher price for a more desired grade or the mid price when blending three close grades. Of course, they risk receiving the price set for the lowest grade leaves in the mix or even a mixed-grade price, much lower, if a grader notes the mixing of grades.

Roy Greene responded to the farmer's tendency to mix grades from the perspective of his experience as a government grader. "There is a tendency [for farmers to not] sort it as carefully as I did when I first went into the grading service. They mix strip and [the graders] denote that with an M and if everything in it is of first quality, first quality X's and first quality C's, you refer to it as MlF. Now a mixed strip group, there are three groups that are

The auctioneer, buyers, and staff who support them. (J. Winston Coleman Kentuckiana Collection, Transylvania University Library)

tied together, are put in the same lot, referred to as M, three closely related groups or it could be two not closely related groups. Now the M on that end means a mixed strip, there are several groups included in that lot. M's on the other end, like V1M or C1M, it means it's mixed in colors."

The daily auctions go to each warehouse on a rotating schedule, beginning at 9:00 in the morning. As Roy Greene described, "You don't sell at the same time each day. The first to sell today sells last tomorrow, then he moves up. They rotate it. Everybody would rather have the first sale. They just feel like that's advantageous. If there is a lot of bad tobacco on them and they sell all day and then come to your tobacco, they might not be very interested."

Nell Collins said, "The buyers come along and they go down one aisle after another and they have an auctioneer crying it off and the high bidder gets the basket of tobacco." The auction process is rapid, and the auctioneer and the buyers move along at a steady walk. The auctioneer is assisted by a "starter" who, with hand signals or by yelling it out, informs him of the "starting price" that is three to four dollars over the support price, depending on what the market is doing. Another man may help the auctioneer with spotting bids. There is always a woman with a small clipboard who follows the spotter and records the bid on the form that already includes the seller's name, a three-digit grade code, and grade price. An auctioneer tries to get the price as stated by the starter. If he doesn't, he drops down a dollar at a time until he picks up a bid. Then he tries to go up. After a few days of the market, the auctioneer has a clear idea of what buyers are willing to pay and it goes quickly.

The grade of tobacco determines the price at which bidding may begin, but tobacco of the same grade may bring different returns. A walk through the warehouse floor after a sale reveals many examples of this. Farmers are reluctant to point out flaws in their neighbors' crops, but in a general way they show how the handling and stripping of a grade can make a big difference in the return on a crop.

Farmers often comment that the tobacco is not truly auctioned and that the buyers are not bidding against each other in any real sense. Auctioneers and buyers indicate that this is true, and that applies most clearly where the crop is in short supply and prices would tend to go up. The process is called "splitting the sale." According to Charles Fowlkes, "We auctioneers do decide on the percentage of the tobacco the companies will get, but we usually do that if it's a year where there is real high demand [and] all companies want it. The way we distribute it is we sell it to the companies in relation to what they have bought in years prior. If they bought 30 percent of the sale or 20 percent [that's what they get]. They pretty much bring the same price. That's what we call splitting the sale. If the company bought 30 percent in years prior, we try to give them 30 percent of the tobacco. You have to carry that in your mind. Some people wonder how I can carry twelve or thirteen companies out there and everybody's buying a different percent. It's just experience I guess. We don't get it perfect. Better auctioneers get it pretty well down pat. If at the end of the day we are a little bit off, [we] make it up. You want to be fair to each company. It seems like my job would be easy when everybody is bidding the same thing, say $1.90, and everybody is going down the row and I've got ten dealers bidding $1.90. I've got to make a decision. That's when it becomes really hard mentally because one company is getting 20 percent, the other 30 percent, this one about 18 percent, one 12 percent, and one 4 percent. You've got to carry all that stuff in your head. It's a little bit hectic. I'd rather for it to be selling at different prices."

The buyer's perspective in this is expressed by Clinton Bramel, who stressed the importance of having a good relationship with the auctioneer and the warehouseman. "You need to have a warehouseman on your side. In a year where we have an allotment, like if you have a season where the crop is short, where the crop is of poor quality, sometimes there is such a demand for it that it all brings the same price and when it does that, the auctioneer just divides the tobacco up amongst the companies. Everybody may be bidding on each pile of tobacco and bidding the same amount of money, but they usually base it on the amount of tobacco the company has bought in previous years. If you are a larger company, they will sell you more tobacco. Smaller companies get less. The warehouseman, if he likes you real well, if you get along with him real well, even though you are a smaller company you might get a better split. They usually base it on what your company has bought in previous years. Take for instance if there are ten piles of tobacco in a particular crop and there are ten buyers. The largest buyer might get three piles out of that. If the warehouseman likes you, you might get a pile. If he didn't, you might not get any. The crop might dry up before you'd get one to suit your grade."

The actual price paid when the participants are splitting the crop is determined by the large buyers such as R.J. Reynolds. Bramel described this. "Usually [you] let domestics set the price. They are much larger companies. Say the market starts out at a $1.90 a pound on a particular type and R.J. Reynolds buys the first pile on the first day of the sale. Then the other dealers and other companies will follow at the same price. If no one tops that dollar, it usually establishes the market. In a short crop year, it would be foolish for a dealer to jump the price up to say $1.92 because he is not going to outbid someone as large as R.J. Reynolds because if they need the tobacco they are going to have it, too. If there are ten buyers in the line, they'll all be bidding $1.90 and it will be allocated."

When the crop is plentiful, buyers behave in a way that helps keep the price stable as the buyers fill their quotas. As Clinton Bramel described it, "Last year there was an extremely large crop, it was good quality, and the companies knew there was more [tobacco] than they could possibly use. So they sort of spaced themselves out for the whole season. Some of the dealers would have their buyers buy a certain amount per day so they would spread themselves out over the whole market rather than buy it all up in three weeks and then just quit. That would make the market drop a few cents which would hurt the farmer. The companies are pretty good about trying to help the farmer. They try to buy the different types of tobacco during the whole season."

The buyers are good judges of tobacco. As they buy it, they assign a company grade that is recorded by a company representative. The company representative marks the tobacco for "moving it off sale." There may also be a company supervisor who monitors the kind of tobacco purchased to keep track of the company's position. Not all companies have buyers of their own. Instead, other firms do their buying for them. Southwestern buys for Philip Morris, for example.

Roy Greene pointed out women who were following the sale. "They look at [the tobacco], and if they don't see anything wrong with it, they put the ribbon on it. Each company has a different color." The ribbons marking the purchases are useful to the warehouseman when he "breaks the sale," or segregates the tobacco into separate lots for shipment to the companies.

If something is found to be wrong with the tobacco, it is rejected and the lots must be resold. Farmers can also reject the sale if they are not satisfied with the price. This needs to be done within a half hour. Sometimes acquaintances are instructed to reject a sale if the farmer's price goals have not been met. Rejection is accomplished by tearing the tag. When a bid is accepted, the tag goes to the office.

The tags are collected by the warehousemen. Dan Grady explained this procedure. "They take them and put them on the records, see, and they take this tobacco, and the people up at the office send them a bill for it, and then they send them a check. It all goes just like buying a sack of flour used to years and years ago." The farmers are given their checks shortly after the sale is finished. Sometimes the checks are brought out to the waiting room by the warehouse owner and passed out. The weight of their sale is subtracted from the total market quota on their marketing card.

11
Burley Tobacco and Its Transformations

However buffeted by market and political forces, medical and legal threats, the four hundred-year-old craft of tobacco production is still in place. The basic outlines of the crop production system used in the Virginia Tidewater in the seventeenth century and among Native American horticulturists even earlier can still be seen in the practices of contemporary tobacco farmers (Carman 1939, Green 1965, Isaac 1982, 22-27, Smyth 1784). At the same time tobacco knowledge and practices have changed through time. Yields have increased. Market prices have stabilized. Labor has been saved. Moreover, as all are aware, the end of the twentieth century has brought some important challenges. These include new understandings about the effect of tobacco use on health, development of offshore sources of burley tobacco, reduction in the amount of domestic tobacco used in cigarettes because of technical innovations in manufacturing, and the increasingly difficult farm labor supply problem. These factors influence tobacco producers and production modes both directly and through the uncertainty they cause.

This uncertainty curtails technical development and innovation, inhibits the development of research facilities and tobacco researcher careers, and raises questions about the very future of the farmers, families, and communities now raising burley tobacco. While it is still labor-intensive, still dominated by family farms, and, above all, still requires complex, hard-won knowledge to produce a crop "in good order," the really important questions about the future of tobacco are as much cultural as economic and technical.

We started this book with a discussion of factors that made burley tobacco culture uniquely elaborate. Here, as we conclude, we address transformations in the knowledge, beliefs, and practices that make up tobacco culture.

Starting at the most concrete level, we can say the tobacco itself has been modified, the machinery and buildings used to deal with it have been improved, and the agronomic management practices of field preparation, field rotation, fertilization, and cultivation have been transformed. Marketing of tobacco is different now than in the past. And the social and political dimensions of tobacco culture have changed. The social basis of tobacco labor has been transformed. The political influence of tobacco people is decreasing. At the symbolic level, tobacco has become increasingly contested.

The factors that lend stability to this particular commodity culture represent limits to the cultural transformation of burley tobacco. The nature of the tobacco plant limits technical innovations in its handling. The modern tobacco plant may be resistant to certain diseases, more responsive to nitrogen, and taller, but it is still tobacco. Hard to handle and easy to damage.

While it may seem rhetorical to say that the nature of an agricultural plant structures its production system, this is especially true in the case of tobacco. Burley, like other kinds of tobacco, is one of the most difficult agricultural commodities to mechanize. The newly harvested leaf can easily be bruised or broken off the stalk and lost. George Duncan said, "You can't bounce it around without breakage and damage." Tobacco cannot be made to flow like grain or potatoes. Each plant must be handled carefully to preserve its value and high potential for return on the farmer's investment. These conditions make use of conveyer systems difficult.

Burley tobacco culture is different from flue-cured tobacco culture. From the perspective of production technology, the key difference is burley's relatively lower sugar content, which re-

Burley Tobacco and Its Transformations 181

duces the tendency of the tobacco to rot and allows a number of its characteristic production practices. Burley can be harvested plant by plant; whereas flue-cured must be primed, or harvested leaf by leaf. Burley also is slowly air-cured rather than cured quickly through the application of heat.

Although there are changes in barns and curing technology, investment in the traditional on-farm, post-harvest curing infrastructure is a major constraint to innovation. Farmers have a huge investment in sticks and barns. The value of sticks alone is high. George Duncan noted, "I've been computing how many sticks there are in Kentucky that are used for tobacco harvest each year. I've roughly estimated that if each stick is worth a dollar, it's probably almost equivalent to the state budget each year for the number of sticks that are in Kentucky. So, if you can't use them as they are, you're talking about a tremendous investment to replace them." The existing tobacco curing barns represent an investment much larger than the burley belt's inventory of sticks. On economic grounds alone, it is difficult to develop a system of curing based on something other than the conventional barn. George Duncan summarized the situation. "If anything new comes along, it either must be cheap enough that the farmer can afford to walk away from the barn, or it must be adaptable to the barn in a retrofit or a reconversion or remodeling method."

Another major constraint to mechanization is the small size of individual tobacco farm marketing units. There are more than 120,000 marketing quotas in burley tobacco. According to George Duncan, "Eighty-seven percent of the growers in Kentucky are under two acres." The tobacco income of individual farms is not great enough to allow investment in the harvest machinery. Machinery designers have not been able to design machines that could be economically sensible for small producers. Workable designs apparently have been developed that would be feasible for fifteen- to twenty-acre tobacco operations.

Because burley tobacco production is carried out on numerous small-scale family-owned farms rather than large-scale corporate farms, the special knowledge associated with tobacco production culture is widely distributed in the region's communities. Further, community welfare and tobacco are closely related. This can be expressed in unexpected places. One of us observed, while attending a 1996 election-eve rally at the University of Kentucky at which President Clinton and Hillary Rodham Clinton appeared, a man carrying a placard that conveyed this theme. The sign, which he had folded up to get past party operatives and then held up for the crowd, was a retort to Hillary's slogan "It Takes a Village to Raise a Child." His sign exclaimed, "No Tobacco—No Village" in response to the anti-tobacco aspects of President Clinton's political agenda.

The long-term importance of tobacco in American economic life still has a stabilizing impact on the politics of tobacco. These days, with concern about export and economic globalization, American cigarette brands are recognized, purchased, and smoked all over the world. We have seen the familiar red and white Marlboro box throughout Europe and Asia. Export sales of cigarettes and processed tobacco are large. The consumption of cigarettes is increasing worldwide. A major portion of those cigarettes are made in the style of the American blended cigarette. This tends to stabilize the commodity in a number of ways.

Yet with all these stabilizing influences at work, the culture of tobacco production continues to change. Advances in knowledge of agronomy and plant breeding, declines in the availability of local labor, alteration of government programs and public attitudes toward tobacco—all these have both direct and indirect consequences. For a clearer understanding of how tobacco production is affected, it will be helpful to examine some of the processes of change in more detail.

Intensification. A dominant theme in the history of burley tobacco is the intensification of production. That is, farmers worked to produce more tobacco from a given unit of land by better management and more inputs. Intensified production cannot be sustained without the addition of nutrients. Before the availability of chemical fertilizers, long fallow periods were required in the rotation system. With the advent of chemical fertilizers, rotations became shorter and involved much less fallow land. Older farmers remember the process of "grubbing out a field" that had grown up in brush after its capacity to grow a decent crop of tobacco diminished. Cropping cycles are much shorter now, true fallowing is a distant memory, and "new ground" is, of course, something to read about in history books. As anthropologists, we found the early pattern of rotation especially interesting because it reminded us of the long fallow systems used by slash-and-burn horticulturists often studied by anthropologists.

These days it is possible to plant tobacco on the same plot year after year. This requires disease-resistant plants and plant breeding programs. High fertilization has increased soil acidity, which leads to increases in the plant uptake of manganese to the point where there are toxic effects on the plant. According to Gary Palmer, University of Kentucky agronomist, "This has consistently been one of the most damaging problems in tobacco production."

The shift from an acreage-based to a weight-based market quota system in the early 1970s had a number of effects, including a general reduction of intensity of production. There was less emphasis on yield increases through improved plants and heavy applications of fertilizer. According to a recent study, "Prior to the adoption of the poundage program in 1971, producers had an incentive to maximize yields for a given acreage constraint. This observation coupled with technological advances resulted in yields increasing at a rapid pace during the 1960s. Since 1971 burley yields have been trending downwards as growers have substituted land for various yield enhancing inputs" (Snell et al. 1991, 9).

Risk reduction. People who are not farmers tend to see technological innovations in agriculture as aimed toward saving labor or increasing yields. In addition to these important goals, farmers and research scientists have developed technology that resulted in risk reduction and increases in managerial control. A case in point is the shift away from earlier bed preparation techniques. Gassing, rather than burning beds, reduces risks by increasing the weed-killing effectiveness of bed preparation practices. Burning beds usually necessitated additional plant bed weeding that came during a time of high labor demand for other crops. Gassed beds have fewer weeds, and the period of time within which gassing can be done is longer than the time in which beds can be burned. This means greater flexibility for the farmer, allowing him or her to more effectively shift tasks among activities.

The use of transplants has some of the same effects as bed gassing. Using transplants allows a significant reduction in time management problems. Under these circumstances, farmers do not have to deal with the complex task of coordinating the plant beds and the planting. An example of a managerial "crunch" associated with coordination in this realm is the effect of heavy rains, which accelerate the growth of plant beds while delaying planting.

Perhaps the best examples of beneficial increases in managerial control derived from technical innovation are those that have come from planting machines. Most older farmers experienced a technical revolution in this aspect of the production process that is similar to the change from farming in the Neolithic Age to farming in the Industrial Age. While we can speak of the new technology allowing time shifting and increased productivity, we must emphasize that the earliest process of "pegging it in" was exhausting drudgery. Visualize using your fingers or a simple hand-carved dibble to perforate the moist soil plant after plant. Keeping the important labor-saving aspects of setters in mind, the new practices increased managerial control and reduced risk in significant ways. Pegging it in required a season, as they say. The soil had to have the right moisture conditions for the plant to survive. If there wasn't rain prior to planting, the transplants could not be planted. The modern transplanter creates its own season by providing water to the plant. This allows the plants to get established even if there is not a season. Ironically the conditions that might be ideal for pegging it in may be problematic for the use of a mechanical transplanter. The changes from the earliest horse-drawn implements to the present designs geared to the use of float plants are little more than elaborations of the basic breakthrough. Mechanical planters allowed farmers to escape the weather.

Reduced costs. Farmers are increasingly sensitive to production costs. Several technical innovations have significantly reduced production costs. The change in the way tobacco is packaged for market is a good example. The century started with the use of hogsheads. The shift to loose-leaf sales of hand-tied tobacco was consistent with the auction system. Although aesthetically pleasing and consistent with careful grading, this system gave way to the much more efficient baling system. Baling reduced farmer cost at a time when labor was becoming increasingly scarce. It also changed other aspects of packaging, in that baling allowed drier tobacco to be stripped, although it seemed to increase the risk of rotting if the tobacco was in high order. This meant there was less waiting for a season once the tobacco was cured. Further, more tobacco could be taken down at one time, once stripping started. Baling has had the effect of reducing warehouse labor costs and the employment of unskilled urban labor.

Many chemicals used in tobacco save labor. The introduction of maleic hydrazine sucker control has been an important source of labor savings and yield improvement. MH entirely eliminated the tedious and time-consuming suckering process. The use of herbicides has reduced the amount of hand-hoeing and cultivation.

Decreased self-sufficiency. More and more tobacco farmers are dependent on others for the materials used in the production process. Perhaps the clearest example of this is the dramatic increase in the use of agricultural chemicals such as methyl bromide, fertilizer, maleic hydrazine, herbicides, fungicides and insecticides, and petroleum-based fuels. That is, farmers have become increasingly dependent on industrially supplied chemicals and fuels for which there is a cash outlay. Social scientists refer to this process as delocalization.

Nowhere in the process of tobacco production is the use of agricultural chemicals more apparent than in the period between planting and harvest. The standing crop is valuable because a great deal already has been invested in it. Therefore, farmers use chemicals to protect it. Herbicides have become widely used and have had considerable impact on weed-control practices. The availability of herbicides is linked to mechanization in an interesting way. The fragile nature of the growing tobacco plant limits the use of tractors in cultivation. While tractors are used to cultivate, there is risk that the tractor may damage the standing crop. Using horses, farmers could cultivate more often and later in the season than with tractors. Herbicides permitted the transition from these earlier modes.

Insecticides are also an important part of the crop protection story. Talking to people about their childhood experiences in tobacco makes one sensitive to at least one impact of the use of insecticides. These chemicals have eliminated the often repeated, vivid childhood memory of "worming tobacco." Worming was something children were good at. They all did it. It was a nasty job.

Increased use of agricultural chemicals in tobacco production served as the foundation for an explanation for how tobacco use causes disease which deflects attention from tobacco itself. Some farmers we interviewed suggested that the chemicals used in contemporary tobacco production may be the real health problem. This was often expressed as a question rather than a statement of belief. For example, Wendell Berry wrote, "I would like to know the effect of the residues of agricultural chemicals in the tobacco" (1993, 9).

Not all delocalization consists of chemical use; it also appears in the transplant-production process. Farmers are less self-sufficient in regards to bed preparation and the production of seed and plants than at any time in the past. Most dramatic are the risks of diseased plants from the transplant market and increased cost. While there was risk with plants raised on the farm, farmers could reduce this risk through their own practices.

Increases in managerial control and reductions in labor cost are associated with increases in the scale of tobacco farm operations. Through purchases and rental of farms, the number of larger-scale producers has increased with a corresponding decrease in the smaller-scale producers. More farmers are raising ten, twenty, and thirty acres of tobacco than in the past. Marketing quotas over 100,000 pounds are increasingly apparent. Increasingly, smaller-scale producers will rely heavily on off-farm employment to achieve their family's economic goals.

Changing categories. In the stripping room, farmers try to make the most of their success in production and curing through the application of their knowledge of grading tobacco leaf. One is most aware of tobacco culture in the setting of the stripping room. In the old days, classification was more complex than today. Another oft repeated story from the past refers to how Daddy or Grandpa stripped the tobacco into five, six, seven, even ten grades, reflecting subtle nuances of color, size, and quality as well as stalk position. The pricing system operating in the past must have rewarded this meticulous classification. Indeed some farmers said some tobacco was not purchased at all. This has changed dramatically. Some farmers in some crops will strip into one grade or maybe two or three. When there is a short crop, there is even less inclination to strip the tobacco into many grades because there is a reduction of the price differential between grades. The practical application of the grades changes in reference to market conditions because of the costs associated with using larger numbers of categories. This is problematic because a major dimension of quality is classification. Quality tobacco is "good tobacco, well handled." An attribute that justifies the premium cost of American burley on world markets is the way it is handled. The reduction in careful grading can reduce the competitiveness of American tobacco.

In addition to the impact of market conditions and labor costs, changes in tobacco and cigarette production technology influence the meaning of grades. The notion of what is the most valuable leaf has changed through time. Prior to the development of the blended cigarette, the light leaf at the bottom of the stalk had less value. Some speak of buyers paying for up-stalk leaf, such as lugs and brights, but asking the farmer to throw the flyings in for nothing. The blended cigarette made good use of the light leaf at the bottom of the stalk; therefore, the value of this leaf increased. The development of filtered cigarettes resulted in increased demand for the heavier, more flavorful up-stalk leaf.

Some categories, such as "cutters," are no longer used. This type of leaf is now included in other grades. Cutters were a high quality grade found in the middle of the stalk between the flyings and the leaf. These long leaves were used as wrappers, for which there is no longer a demand. Because there is no current use for this type of leaf, it is now divided into the grades just above and below it on the stalk (the flyings and the lugs).

Changing demography and labor supply. The number of people available for work in tobacco has steadily declined. The rural population declined generally, families got smaller, and farmers got older and less able to work. In the past, farmers swapped labor, relied heavily on family labor, and were more able to hire people. All of these sources of labor are more difficult to obtain now than in the past. Labor swapping became rare because increases in the amount of off-farm employment made it difficult to schedule the shared work. Family labor declined because of the changing family of the increasingly older farmer population. Fifty- and sixty-year-old farmers have fewer children at home, and farmers' children are now more involved in their own extracurricular activities or off-farm employment. Financially, they are better off working at fast food establishments than working on the farm. Perhaps most striking in the inner Bluegrass region of Kentucky is the virtual disappearance of African-American hired hands. All these processes are further complicated by reduction in the skill levels of the labor force for the same reasons that reduced the amount of available labor. The reduction in the use of labor swapping and family labor means laborers must be hired for peak seasons. Farmers have trouble competing with local industry in hiring part-time farm labor as factories provide well-paid, secure jobs. The farmers cannot afford to pay well, employment is seasonal, and farm labor is strenuous and sometimes dangerous. Because of the difficulty in hiring local labor, the number of migrant farm workers in burley tobacco has increased rapidly.

The transformation in labor has important consequences for communities. The social transformation from working alongside a family member or a neighboring fellow farmer or one of his children to hiring and training migrant labor is drastic. The practice of labor swapping increased community solidarity through the creation of networks of mutual aid that resulted in long-term reciprocal relationships between neighbors. Hiring local folks during the peak labor periods of planting, housing, and stripping served to create similar durable, mutually supportive relationships. The loss of community solidarity associated with changes in the way agricultural work is done transforms communities.

Organizing to improve prices and stabilize supply. The development of the tobacco program is a dramatic story of a grass-roots farmer movement struggling fitfully and then, largely because of the destabilization caused by the Great Depression, emerging more or less fully supported by government policy. While early efforts to organize tobacco producers had limited success (Campbell 1993), the early struggles led to the creation and testing of approaches for reducing chronic oversupply and subsequent disinflation in tobacco prices. An important aspect of the successful emergence of this system is the shared interests that existed between tenants, farmers, warehousemen, and the regional elite. As pure producer organizations, early farmer organizations might not have worked. While there appears uncertainty about its future, the program is part of the fabric of the burley tobacco culture. As such, it structures farmer decision making and provides an unambiguous template with which much of the rural population judges the worth of a politician. In an effort to capitalize on this theme, a physician who was running as a Republican against the Kentucky second district congressional representative was reported to have sent a letter to his mailing list of tobacco-involved voters indicating that as a Republican he supported the tobacco program. The ironic chord was elaborated by the fact that Scotty Baesler, his opponent, was the only tobacco farmer in Congress at the time. In central Kentucky, an anti-big government candidate, presumably professionally committed to health, thought it politically necessary to support the tobacco program by trying to depict a tobacco farming New Deal legatee as not the tobacco farmer's right choice.

The tobacco program stabilized the tobacco market, increased farm income, reduced risk, and led to increased land values and tax receipts. All of our informants agreed that the Burley Tobacco Program was necessary to the security of the tobacco farmer. The program's future, however, is uncertain. A question is raised: "Can market quota and price incentives work together to get the supply right?" This is complicated by the increasing availability of tobacco produced outside of America and the essentially flat domestic demand for cigarettes. Export considerations have much more importance in the management decisions associated with the tobacco program. Recent tobacco program management practices, forged of political compromise at the national level, require that cigarettes contain a certain minimum of domestically raised leaf.

A pervasive factor structuring the world of tobacco from the development of tobacco production technology to the training of tobacco researchers is the indefinite future of tobacco. Starting with the surgeon general's report in 1964, tobacco changed from a commodity with a developing future to a commodity in decline. This has many effects, including the potential for the tobacco culture to shift from durable and stable to impermanent. This shift will have a profound effect on tobacco culture. Tobacco is still the most important commodity economically where it is produced. It is difficult to make investments in technology development at various levels in the system because of the uncertainty about tobacco and related problems paying for such an investment. Many things could happen. Domestic demand could drop substantially through taxation or marketing control, and the tobacco program could be deauthorized.

Pressure continues for abolishment of the tobacco program. The rationales used revolve around the implications of government involvement in fostering orderly tobacco economics and policy concern for public health as well as the generalized disenchantment with "big government." Tobacco is being battered by two clubs—one wielded by those attempting to dismantle the New Deal, the other by those who are against smoking. The weakening political position of tobacco can be seen in changes to the tobacco program made by Congress in the 1980s. The no-net-cost program of 1982 and the reduction of price supports in 1985 were steps taken to save the tobacco program from abolition by making political support more palatable. These changes allowed farmers and their political leadership to talk about the tobacco program not costing the taxpayers any money. Many people outside and within the tobacco regions say the government should be out of as many things as possible, particularly anything that supports an industry that produces a health-harming, addictive non-food crop.

The ultimate disconnection—the abolition of the tobacco program—is still a threat. If this were to happen, farm incomes would fall, manufacturing costs would decline, and the value of farm land would decrease (Wagner 1971, 244). This would lead to the consolidation of small land holdings into larger, more mechanized farms, as has happened in flue-cured and cigar tobacco growing areas. The threat to tenant farmers is especially great. Most people agree with Virginia Calk, who said, "Most tenant farmers depend on tobacco alone for their support and if it fails, it would be awful."

Our concern in writing this book was primarily with describing, in the words of tobacco farmers, the process of raising a crop of tobacco, rather than the changing meanings attached to tobacco by society at large because of the health problems associated with tobacco use. This approach reflected our interest in the craft of tobacco farming rather than public health. Yet, upon reflection, one can say meanings are everything. The crucial thing about tobacco today is not the technical details of commodity production, but the meanings that the society at large attaches to its production. It is difficult to find even one technical aspect of the crop that is not influenced by the symbolic transformation of tobacco because of the impact of tobacco on health. The omnibus problem is that the health cost of tobacco use results in increasing regulatory and economic constraints to tobacco use, education to reduce tobacco use, and a general symbolic devaluing of tobacco. This increases economic uncertainty and the political cost of supporting government programs for tobacco. Because of this, the technical realm sees decreases in the amount of public money spent on the development of new technology, decreases in its diffusion through agricultural Extension, and decreases in adoption of technology by farmers.

In response to the economic uncertainty caused by countrywide litigation against them, major tobacco companies, including Philip Morris, RJR Nabisco, B.A.T., Brown and Williamson, and Lorillard, entered into negotiations with attorneys general from nearly forty states. In June 1997 an agreement was signed by the participating parties which would, when approved by Congress and signed by the President, put into effect a complex and expensive program directed at curtailing tobacco use, especially by youth, and punishing tobacco companies while it reduced tobacco company exposure to liability. The companies would pay $368.5 billion over the next twenty-five years to compensate states for their tobacco-related health costs and to create a fund for paying successful individual law suits against them. Funds would also be used to support health research and education programs aimed mostly at youth. The program would require comprehensive changes in how tobacco was marketed. These include new, more prominently displayed warning labels, elimination of self-service, mail, and machine sales, licensing of vendors, limitations in media placements, paraphernalia, and promotional gifts, and restrictions on the appearance, content, and location of advertising. The impact of these changes would be monitored. Tobacco companies will be required to take responsibility for reducing tobacco use among youth. If certain targets of decrease are not reached very large fines must be paid.

The program would establish the Food and Drug Administration as a regulator of cigarettes. Tobacco companies would be required to disclose company records and their ingredients on cigarette packs. Each package would have to include the statement that cigarettes are made as a nicotine delivery device.

The content and meaning of the agreement was intensely debated by health advocates, politicians, and farmers. Health advocates seemed to be concerned about whether the program was tough enough on the companies. Although one health advocate wrote, "the settlement appears to provide the public health advocates with everything realistic on their 'wish list' of tobacco controls—with the exception of vengeance" (Bailey 1997). The post-agreement debate reminded the public that farmers did not participate in the development of the agreement.

It is difficult to predict from the public debate whether and under what circumstances the agreement would be put into effect. Members of Congress appear to resist this legislative agreement that they did not fashion themselves. The criticisms of the tobacco companies by health advocates has not quieted in spite of the scope of the plan. Tobacco companies are not in a position to break things loose. The existence of negotiations themselves communicated the tobacco companies' economic vulnerability. They switched from being tough litigants to compliant negotiators. The farmers and farm communities wonder where they fit in.

The very existence of the agreement has already had important political impacts. During Congressional hearings about the agreement the abolishment of the tobacco program was revisited. The Chairman of the Senate Agriculture Committee, Senator Richard Lugar (R), proposed a buyout of market quota as part of a plan to abolish the tobacco program. Farmers would be free to produce whatever amount of tobacco they would like, wherever they could produce it, without a floor price.

Culturally tobacco has been transformed from a commodity with a rich history and a useful future to a commodity with only an uncertain present. The tobacco production community's main task is adjusting to a culture of impermanence. This is paradoxical. Tobacco is anything but impermanent. Anticipated demand for the 1997 crop will be at a record high. University of Kentucky agricultural economist William Snell said, "It amazes me, given all the challenges this industry is facing, we're talking about an all-time record quota" (Lucke 1997, 1). We think tobacco culture will be with us for some time to come.

Notes

1 Tobacco Culture

1. References to negative health effects of tobacco use appear in very early tobacco literature. The anti-smoking tract *Counterblaste to Tobacco*, written by the English King James I, was published in 1604. Understanding of the epidemiological and genetic evidence is very recent.

2 Tobacco Ground

1. When prepared as a powder rather than granules, this is used as an explosive in mining and construction. This requires mixing with fuel oil and use of a detonator.

3 Tobacco Labor

1. The meals served to us possibly were more elaborate than those served when we were not there.
2. Such workers had been employed in harvesting strawberries for the process market until Kentucky strawberry production declined.

4 The Tobacco Program

1. The average support price was $1.74 in 1996 and $1.72 in 1995. The highest support price in 1996 was $1.81.
2. The group was formally named Burley District Branch of the American Society of Equity Department of Tobacco Growers. It was always called Burley Tobacco Society (Bleidt 1932, 25).
3. The Burley Tobacco Society was related to and modeled after the Society for Equity, a national farmers' organization. In earlier times, local newspapers often printed announcements of meetings of the Society for Equity.
4. Following the 1982 no-net cost legislation, net profits from these sales were put in the "no-net-cost" fund.
5. The name of the ASCS was changed to the Farm Service Agency (FSA) in 1995. Everybody seems to still call it ASCS. We have even run into

Notes

farmers who referred to the ASCS office by its original New Deal nickname: the "Triple A" (or Agricultural Adjustment Administration office).

6. This was reduced to 103 percent in 1985.

5 Sowing the Beds

1. The sacks are used to bundle transplants. The plants are arranged somewhat like a sheaf of wheat, and the sacking is used something like a belt around the bundle. The nail is used like a straight pin to fasten the sacking.

6 Setting the Plants

1. For example, the Speedling Company sells trays of 200-, 242-, 288-, and 392-plant capacity. Lower-capacity trays have larger soil capacity, which reduces daily maintenance and risk because plants in larger cells can live longer under stress.

7 Cultivating and Topping

1. Nabam is a fungicide marketed as Dithane D14 or Parzate Liquid Fungicide.

8 Cutting, Housing, and Curing

1. Interlocking, notched logs were stacked to make a simple, more or less square structure occasionally used for tobacco.

2. The technical description of the curing process is based on Garner, 1946. See Chapter 20 "Curing, Fermentation, and Aging of Tobacco."

3. Tobacco balers use mechanical or pneumatic compression. Systems that use hydraulic fluid are not used around tobacco.

9 The Stripping Room

1. The tie leaf is used to wrap the top of the individual bundle. The leaf must be sound, sufficiently long, and attractive.

2. This may be termed a basket charge, although baskets are no longer used.

10 On the Floor

1. The difference in percentage of water content between "dry" and "wet" tobacco is small. Tobacco that has 14 percent moisture content feels dry. Tobacco that has 22 percent moisture content feels wet.

References Cited

Anschel, Kurt R., and Russell H. Brannon. 1972. Impact of Declining Demand for Tobacco on the Economy of Kentucky. In *Social and Economic Issues Confronting the Tobacco Industry in the Seventies*. S. Frank Bordeaux Jr. and Russell H. Brannon, eds. Lexington, Ky.: University of Kentucky.

Atkinson, W.O. et al. 1981. *Tobacco Handbook*. ID-45 Cooperative Extension Service, College of Agriculture, University of Kentucky.

Axton, William F. 1975. *Tobacco and Kentucky*. Lexington, Ky.: Univ. Press of Kentucky.

Bailey, William J. 1997. Tobacco Settlement at a Glance. Indiana Prevention Resource Center at Indiana University [http://www.drugs.indiana.edu/druginfo/settlement.htm].

Baldwin, Leland D. 1941. *The Keelboat Age on Western Waters*. Pittsburgh: Univ. of Pittsburgh Press.

Begley, Sharon. 1996. The Cancer Killer: It's a cell's most elegant defender, a gene called p53. It stops tumors before they grow. But if damaged, it is involved in 60 percent of cancers. *Newsweek* December 23, 42-47.

Benson, Fred J., Steve Isaacs, Richard L. Trimble. 1993. Field Crop Enterprise Cost and Return Estimates for Kentucky - 1993. Agricultural Economics, Extension No. 55. Lexington, Ky.: College of Agriculture, University of Kentucky.

Berry, John M. Jr. 1993. The History of the Tobacco Program. In *The Tobacco Church: A Manual for Congregational Leaders*. Bennett D. Poage, ed. Richmond, Ky.: The Christian Church (Disciples of Christ) in Kentucky.

Bishop, Jerry E., and Milo Geyelin. 1996. Researchers Show How Smoking Causes Cancer. *Wall Street Journal* October 18, 1996: B1-B2.

Bleidt, Helen Morris. 1932. A History of the Development of Growers' Organizations and Tobacco Pools in the Burley District of Kentucky, 1902-1927. Unpublished Thesis, University of Kentucky.

References Cited

Boylan, B.V. Brandt, J. Muehlbauer, M. Auslander, C. Spurlock, and R. Finger. 1993. Green Tobacco Sickness in Tobacco Harvesters - Kentucky, 1992. *Journ. of the Amer. Medical Assoc.* 269 (21): 1,993, 2,722, 2,742.

Bradford, L.J. 1873. White Tobacco. *Daily Kentucky Yeoman* (Frankfort) 20(77): 3. [Original in the *Falmouth Independent*. This is from a typescript "set" into a pamphlet titled "A Weed from the Wilderness." Tom Middleton, Eminence, Ky., at the State Historical Society.]

Breen, T.H. 1985. *Tobacco Culture: The Mentality of the Great Tidewater Planters on the Eve of Revolution.* Princeton, N. J.: Princeton Univ. Press.

Bridenbaugh, Carl. 1980. *Jamestown: 1544-1699.* New York: Oxford Univ. Press.

Campbell, Tracy. 1993. *The Politics of Despair: Power and Resistance in the Tobacco Wars.* Lexington, Ky.: Univ. Press of Kentucky.

Carlton, Chad and Bill Estep. 1997. Growing Uncertainty. Lexington *Herald-Leader* September 21, 1997: A11.

Carr, Tinnie. Letter. Lexington *Herald-Leader*. November 21, 1981.

Census of Agriculture (1992). *Geographic Area Series. Kentucky State and County Data.* Washington, D. C.: Superintendent of Documents, U.S. Government Printing Office.

Clark, Carl M., and Wilmer Browning. 1953. *Organization of the Looseleaf Tobacco Auction Market.* Lexington, Ky.: Kentucky Agricultural Experiment Station, University of Kentucky.

Clark, Thomas D. 1960. *A History of Kentucky.* Lexington, Ky.: John Bradford Press.

Cunningham, Bill. *On Bended Knees: The Night Rider Story.* Nashville, Tenn.: McClanahan Publishing House.

Daniel, P. 1985. *Breaking the Land: The Transformation of Cotton, Tobacco and Rice Cultures Since 1880.* Chicago, Ill.: Univ. of Illinois Press.

Denissenko, M.F. Annie Pao. Moon-shong Tang, Gerd. P. Pfeifer. 1996. Preferential Formation of Benzo[a]pyrene Adducts at Lung Cancer Mutational Hotspots in *P53*. *Science* 274(5286): 430.

Durden, Robert F. 1975. *The Dukes of Durham, 1865-1929.* Durham, N. C. Duke Univ. Press.

Earle, Carville V. 1975. *The Evolution of a Tidewater Settlement System: All Hallow's Parish, Maryland, 1650-1783.* University of Chicago, Department of Geography, Research Paper No. 170.

Eastwood, Susan C. 1989. Sowing Down the Ground: An Ethnographic Description of Tobacco Farming in Montgomery County, Kentucky. Master's thesis, University of Kentucky.

Ellis, William E. 1982. Robert Worth Bingham and the Crisis of Cooperative Marketing in the Twenties. *Agricultural History* 56.

Fertig, Barbara C. 1986. The Tobacco Tradition in Southern Maryland. *New Jersey Folklife* 11: 8-12.

Funkhouser, W.D. and W.S. Webb. 1928. *Ancient Life in Kentucky*. Frankfort, Ky.: Kentucky Geological Society.

Gage, Charles E. 1933. *American Tobacco Types, Uses, and Markets*. Circular 249. Washington, D. C.: U.S. Department of Agriculture.

Garner, Wightman W. 1946. *The Production of Tobacco*. Philadelphia: Blakiston.

Gehlbach, S.H., L.D. Perry, J.J. Langone, H. Van Vanakis, W.A. Williams, J.I. Freeman, and L.V. Peta. Nicotine Absorption by Workers Harvesting Green Tobacco. *Lancet* 1(7905): 478-480.

Greene, Randall Elisha. 1994. The History and Culture of U.S. Tobacco Growers: An Overview. In *The Tobacco Church II: A Manual for Congregational Leaders*. Bennett D. Poage, ed. Richmond, Ky.: The Christian Church (Disciples of Christ) in Kentucky.

Hart, John Fraser, and Eugene Cotton Mather. 1961. The Character of Tobacco Barns and Their Role in the Tobacco Economy of the United States. *Annals*, Association of American Geographers. 51: 274-293.

Hunt, Albert R. 1997. Waiting for Clinton to Lead. *Wall Street Journal*. Setember 18, 1997: A15.

Irons, Peter H. 1982. *The New Deal Lawyers*. Princeton, N. J.: Princeton Univ. Press.

Isaac, Rhys. 1982. *The Transformation of Virginia, 1740-1790*. New York: W.W. Norton.

Jacobstein, Meyer. 1907. *Tobacco Industry in the United States*. New York: Columbia Univ. Press.

Jahn, Raymond. 1954. *Tobacco Dictionary*. New York: New York Philosophical Library.

Johnson, Paul R. 1984. *The Economics of the Tobacco Industry*. New York: Praeger.

Jones, Hugh. 1956. *The Present State of Virginia, from Whence Is Inferred a Short View of Maryland and North Carolina*. Richard L. Morton, ed. Chapel Hill, N. C.: Univ. of North Carolina Press.

Jones, Robert Leslie. 1983. *History of Agriculture in Ohio to 1880*. Kent, Ohio: Kent State Univ. Press.

Kroll, Harry Harrison. 1965. *Riders in the Night*. Philadelphia, Penn.: Univ. of Pennsylvania Press.

Main, Gloria L. 1982. *Tobacco Colony, Life in Early Maryland, 1650-1720*. Princeton, N. J.: Princeton University Press.

McGrath, Sally V., and Patricia J. McGuire. 1992. *The Money Crop: Tobacco Culture in Calvert County, Maryland*. Crownsville, Md.: Maryland Historical and Cultural Publications.

References Cited

Miller, John G. 1936. *The Black Patch War.* Chapel Hill, N. C.: Univ. of North Carolina Press.
Miller, Robert D., Richard A. Hensley, and Uel D. Wilhoit. 1994. *Instructions for Building Poor Boy Tobacco Seeders.* Research Report 94-04. Univ. of Tennessee, Agricultural Experiment Station.
Murray-Wooley, Carolyn, and Karl Raitz. 1992. *Rock Fences of the Bluegrass.* Lexington, Ky.: Univ. Press of Kentucky.
Nall, James O. 1939. *The Tobacco Night Riders of Kentucky and Tennessee.* Louisville, Ky.: Standard Press.
O'Malley, Nancy. 1989. *Searching for Boonesborough.* Lexington, Ky.: Program for Cultural Resource Assessment.
Osbourne, Kevin. 1990. Tobacco Road, Warehouse Industry Generates Millions for Local Economy. Lexington *Herald-Leader* November 19, 1990: D1, D8.
―――. 1991. Space, Labor Shortage Cited in Slow Sale of Burley Quotas. Lexington *Herald-Leader* June 5th, 1991: B4.
Raitz, Karl B. 1995. Tobacco Barns and Sheds. In *Barns of the Midwest.* Allen G. Noble and Hubert G.H. Wilhelm, eds. Athens, Ohio: Ohio Univ. Press. [125-144]
Redhead, C. Stephen, and Richard E. Rowberg. 1995. *CRS Report for Congress: Environmental Tobacco Smoke and Lung Cancer Risk.* Washington, D. C. Congressional Research Service, Library of Congress.
Robertson County *Times Democrat* June 5, 1924; October 18, 1928; September 15, 1932.
Rosenberg, Gil. 1992. They Were Just Like Family: Framing the Introduction of Hispanic Farm Workers into the Kentucky Tobacco Harvest. Master's thesis, Univ. of Kentucky.
Rosenberg, Gil. 1994. Hispanic Migrant Farmworkers in Kentucky: County Extension Agent Survey. Lexington, Ky.: Univ. of Kentucky, College of Agriculture, Department of Sociology.
Rosenberg, Gil, and C. Milton Coughenour. 1990. The Farm Labor Situation in Kentucky: The Opinions of Farmers and Community Members. *Community Issues.* Univ. of Kentucky, College of Agriculture, Department of Sociology 11(1): 1-9.
Saloutos, Theodore. 1939. The American Society of Equity in Kentucky: A Recent Attempt in Agrarian Reform. *Journal of Southern History* 5: 347-361.
―――. 1960. *Farmer Movements in the South, 1865-1933.* Lincoln, Neb.: Univ. of Nebraska Press.
Shuffett, D. Milton. 1986. Tobacco's Economic Impact in Kentucky. In *Tobacco in Kentucky.* Smiley, et al. eds. Lexington, Ky.: Univ. of Kentucky Cooperative Extension Service.

Smiley et al., eds. 1986. *Tobacco in Kentucky*. Lexington, Ky.: Univ. of Kentucky Cooperative Extension Service.
Smith, J. Allan. 1981. *College of Agriculture, University of Kentucky*. Lexington, Ky.: Agricultural Experiment Station.
Snell, William M. 1993. The Future of Tobacco. In *The Tobacco Church: A Manual for Congregational Leaders*. Bennett D. Poage, ed. Richmond, Ky.: The Christian Church (Disciples of Christ) in Kentucky.
Snell, William M. 1996. The U.S. Tobacco Program: How It Works and Who Pays for It (AEC-82). Lexington, Ky.: Cooperative Extension Service, Univ. of Kentucky.
Snell, William M., D. Milton Shuffett, Orlando D. Chambers, and Perry J. Nutt. 1991. *Long Run Impacts of the Tobacco Improvement Act of 1985, Executive Summary*. Lexington, Ky.: Department of Agricultural Economics, College of Agriculture, Univ. of Kentucky.
Steed, Virgil S. 1947. *Kentucky Tobacco Patch*. Indianapolis: Bobbs-Merril.
Tatham, William. 1800. *An Historical and Practical Essay on the Culture and Commerce of Tobacco*. London.
Tennant, Richard B. 1971. *The American Cigarette Industry: A Study in Economic Analysis and Public Policy*. Archon Books.
Tilley, Nannie M. 1985. *The R.J. Reynolds Tobacco Company*. Chapel Hill, N. C.: Univ. of North Carolina Press.
U.S. Department of Agriculture. 1981. Tobacco Outlook and Situation. *USDA Economic and Statistical Service Quarterly*. March.
——. 1982. World Tobacco Situation. Foreign Agricultural Circular 9-82, USDA. Washington, D. C. U.S. Government Printing Office.
——. 1984. Tobacco Outlook and Situation Report. USDA Economic Research Service. Washington, D. C. U.S. Government Printing Office.
U.S. Public Health Service. 1964. *Smoking and Health: Report of the Advisory Committee to the Surgeon General*. USPHS Publication 1103. Washington, D. C.
van Willigen, John. 1989. *Gettin' Some Age on Me. Social Organization of Older People in a Rural American Community*. Lexington, Ky.: Univ. Press of Kentucky.
Wagner, Susan. 1971. *Cigarette Country: Tobacco in American History and Politics*. New York: Praeger Publishers.
Warren, Robert Penn. 1939. *Night Riders*. New York: Random House.
Weizenecker, Richard, and William B. Deal. 1970. Tobacco Cropper's Sickness. *Journal of Florida Medical Association* 57(12): 13-14.
Winkler, John K. 1942. *Tobacco Tycoon: The Story of James Buchanan Duke*. New York: Random House.

Index

Photographs are in bold.

addiction: nicotine, 6-7. *See also* health problems; nicotine
aesthetics, 86; stripping, 148; packaging, 156, 158; hand-tied tobacco, 185. *See also* pride; tobacco culture
Agricultural Adjustment Act: parity price, concept of, 52; free-rider problem, 53; *United States v. Butler*, 53. *See also* market price; New Deal; tobacco program
Agricultural Adjustment Act Committee: and fertilizer availability, 22; and ASCS, 194-95 n 4.5
agricultural intensification, 182-83; and fertilizer use, 18. *See also* fertilizer; technical innovation
Agricultural Stabilization and Conservation Committee. *See* ASCS
agrochemicals, 102, 185-86; as health risk, 115, 186; and delocalization, 185-86. *See also* fungicides; herbicides; pesticides
air-cured tobacco, 180-81; color, 12-13; in USDA classification system, 12-13; type, 13; Type 31, 13. *See also* white burley tobacco
alternative crops, 1-3; strawberries, 2, 194 n 3.2; bell peppers, 2-3. *See also* crops, other; livestock
American Society of Equity: Everitt, James A., 43
American Tobacco Company, 39-41; the "trust," 39-40; and Bonsack cigarette manufacturing machines, 40. *See also* Duke, James B.; farmers' marketing organizations
ASCS (Agricultural Stabilization and Conservation Service), 19-20; administration of tobacco program, 55; Farmers' Committee, 55; marketing quota, 55, 59; Farm Service Agency, 194. *See also* tobacco program
Atkinson, W.O. "Ted": MH testing, 114
Augusta, Ky., 11-12

Axton, W.F.: *Tobacco and Kentucky*, viii

Baesler, Rep. Scotty, 189
baling, 142-49, 151-54, **152**, 155, 156-59; running them down, 151; process of, 151-52, 153; bales, 152-53; development of, 153-57; price supports, 154; promotion of, 157; advantages of, 157-58, 159; labor, 157-58; market price, 158, 159; support price, 158; objections to, 158-59; storage and handling, 159. *See also* marketing; packaging; stripping room equipment
Barkley, George: origin of white burley tobacco, 11-12
barn, 181; repairs, 81; and curing, 128; design, 128; structure, 128-34, **129**; barn posts, 129; cross ties, 129; tarring with creosote, 129-30; center, 130; sheds, 130; shutters, 130, 136-37; stripping room, 130; vents, 130, 133, 136-37; bents, 130-31, 140; spatial dimensions, 130-31; tiers, 130-31; log-pen type, 131, 195 n 8.1; rafters, 131; tier rails, 131; barn raising, 132; construction tools, 132; construction, 132-33; mortise and tenon type, 132-33; newer barn design, 136; placement, 138; space, 159. *See also* barn space; curing; stripping; tobacco program
barn space, 163; and carry over of poundage, 59-60; double-barning, 135; full, 135; leasing quota, 135; renting, 135. *See also* barn; housing
Bell, Edna: weeding the beds, 80; worming by hand, 107-8; MH sucker control, 115
Bell, Reynolds: tobacco receiving point, 41; purpose of burning the beds, 67, 70; field preparation, 83; cultivation, 102; topping, 112; shank knife, 119; coking, 138-39
benzopyrene: as carcinogen, 7-9; and BPDE, 8. *See also* cancer; nicotine

201

Berry, Wendell: burley tobacco production, care required, 5; effect of agrochemical residues on tobacco, 186
Bingham, Robert W.: Burley Tobacco Growers Cooperative Association, 47-48
Black Patch region: Dark-fired Tobacco District Planters' Protective Association, 41
Bonsack cigarette manufacturing machines: and American Tobacco Company, 40
Boonesborough, Ky., 10-11; tobacco production, early, 10; warehouses, early, 10-11
Bortner, C.E.: MH testing, 114
Bracken County, Ky.: origin of white burley tobacco, 11-12
Bradford, L.J.: description of white burley tobacco, 12; origin of white burley tobacco, 12
Bramel, Clinton: splitting the sale, 176-77
Bridges, Jimmy: economic importance of tobacco, 1; tobacco harvesting machines, 27; pool price, 39; leasing in poundage, 60; sowing the beds, 74; grades, 146; baling, 152-53; government grading, 171; color, 172
Brown, Roy: labor, swapping, 29
Brown County, Ohio: origin of white burley tobacco, 11-12
burley belt, viii; regional tobacco concentration, 3
burley tobacco: economic importance in central Kentucky, 1-4; future, 3; color, 12; historical tradition, 13; soils, 13; varieties, local, 71; varieties, improved, 71-72. *See also* red burley tobacco; white burley tobacco
Burley Tobacco Growers Cooperative Association, 61; pooling, 38-39, 43, 47-48; Barker, Ralph, 47; Baruch, Bernard, 47; Cantrill, J.C., 47; Sapiro, Aaron, 47; Stone, Jim, 47; Bingham, Robert W., 47-48; *Burley Tobacco Grower*, **48, 50**; accomplishments, 48-49; draw payments, 48-49; free riders, 49-50, 50; reasons for failure, 49-50; taking tobacco "under loan," 54; administration of tobacco program pool, 55. *See also* farmers' marketing organizations; tobacco program
Burley Tobacco Program, 38-39, 51-63. *See also* tobacco program
Burley Tobacco Society: Henry County Union, 43; pooling the crop, 43-44; Lebus, Clarence, 44; cut-out of 1908, 44-45; Night Riders, 44-45; and Society of Equity, 194 n 4.2, 194 n 4.3. *See also* farmers' marketing organizations

Calk, Virginia: marketing quota, poundage system, 59; baling and labor, 153; market price for bales, 159; weight, 162; economic importance of tobacco, 191
Camel brand cigarette: use of burley tobacco in, 13
Canada: adoption of baling, 156
cancer: smoking-related, 7-9; benzopyrene, 7-9; carcinogens, 7-9; and public discourse, 8-9. *See also* health problems; nicotine; smoking
carcinogens, 7-9; benzopyrene, 7-9. *See also* cancer; health problems; nicotine; smoking
Carmack, Brad: mixed grades, 148
Carr, Tinnie G.: baling and aesthetics, 158
Carraco, Paul, ix; clearing, 15-16; crop placement on hillsides, 16-17; manure, 20; WWI, effects on burley market, 46-47; farmers' marketing organizations, 47; draw payments, 49; dump houses, 49-50; tobacco program, 54-55; steaming the beds, 67-69; gassing the beds, 69; improved varieties, 72; cottons, 79; field preparation, 81-82; seedbed management, off-season, 82; pegging it in, 86; hand setters, 88; setting with wheel-type setter, 94; seedlings, from neighbors, 96-97; black shank, 109-10; dropping sticks from a high-boy sprayer, 118; splitting tobacco, 119-20; spearing tobacco, 121-22; wilting, 125-26; houseburn, 137-38; other farm work, 140-41; stripping room preparation, 140-41; case, 143-44; marketing, 162-63; making a basket of hands, 166; handling bales at warehouse, 166-67; poor quality leaf, 170

Index

case, 143-44, 158; case getting, 143-44; high case, 144, 162; high order, 144, 162; hot tobacco, 144. *See also* curing; season; stripping; tobacco leaf
cattle: as alternative to tobacco, 1-2; dairying, 1-2; livestock, 1-2; used for clearing, 15-16; grazing, 18
change, 180; seedling production, 83, 85, 100; packaging, 149; uncertain future, 179-80, 190, 191-93; tobacco culture, 179-93; marketing, 180; intensification, 182-83; categories, 187-88; demography and labor supply, 188-89. *See also* mechanization; technical innovations; tobacco culture
chewing tobacco products: decreased demand, 13
cigar-binder tobacco: USDA tobacco classification, 12
cigarette: Camel brand, 13; development of blended type, 13; use of burley tobacco in, 13; Marlboro brand, 182; world market, 182; and grading, 187-88
cigar-filler tobacco: USDA tobacco classification, 12
cigar-wrapper tobacco: USDA tobacco classification, 12
classification: as cultural activity, 5; of burley tobacco leaf, 5, 13; USDA classification system for tobacco, 12-13. *See also* farmers' grades; government grades; grading; tobacco culture
Clay, Berle: sharecropping, 34, 36-37; out-migration, 37; acreage-based allotment system, 56-57; burning the beds, 66; cultivation, and herbicide use, 105-6; relationship between warehouse and grower, 169-70
clearing: debris, 15; grubbing out a field, 15, 183; new ground, 15, 183; use of cattle, 15-16
Cleaver, Silas: hired labor, changes in, 30-31; setting with shoe-type setter, 89-90; pesticide, Paris green, 108; using anti-strut compounds, 139-40
Collins, Nell: pulling plants, 95; topping, 110; falling out of the barn, 135-36; curing, case, 143; bulking it down, 144; tobacco auction, process of, 174-75

Collins, Paul: marketing with the "trust," 40-41; Burley Tobacco Society, 44-45; pinhookers, 51; seed processing, on-farm, 70; setting with shoe-type setter, 90-91; cultivation with work animals, 103; stripping, farmer's grades, 147
Collins, W.B.: varieties of tobacco, local, 71; hoeing by hand, 105
color, 5-6; air-cured tobacco, 12-13; white burley, 13; after spraying with MH, 114-15; and curing, 116-17, 137; high color, 117; and grading, 171, 172
Commodity Credit Corporation: loans, 38-39
Cooper, Senator John Sherman: development of poundage system, 56
cottons, 79. *See also* seedbed management, protective covers
cover crops, 17-19, 81; clover, 9, 18; and crop rotation, 17-18; wheat, 17-18, 140, 141; grass, 17-19; vetch, 18; soybeans, 19; rye, 140, 141. *See also* sowing down the ground
crop placement: bottom land, 16; hillsides, 16-17. *See also* soil fertility; topography
crop rotation, 14-19; fallowing, 14-15; corn, 17; grain, 17-18; grass, 17-18; size of farm, 18; fallow periods, 182-83. *See also* cover crops; land use; soil fertility; sowing down the ground
crops, other: grain, 1, 11; as alternative to tobacco, 1-3; hemp, 11; and crop rotation, 17-18; garden crops, 77; corn, 104, 140; hay, 104, 140; soybeans, 140; clover, 169-70; wheat, 169-70; strawberries, 194 n 3.2. *See also* alternative crops
cultivation, 102-7, **103**; disease, 102; lying by, 102, 106; mechanization, 102; pests, 102, 107 (*see also* pest control; pesticides); weeds, 102; with work animals, 102-4, 103; past technology, 102-5; plowing it hard, 104; chopping, 104-5; hoeing by hand, 104-5; herbicide use, 105-6; side-dressing, 105-6; timing, 106; weather conditions, 106-7; worming tobacco, 186. *See also* cultivation equipment

204 Index

cultivation equipment: tractors, 102; five-shovel cultivator, 103; rastus plow, 104; hoe, 104-5. *See also* field preparation equipment
cultural ecology, x
curing, 5-6, 136-37, **139**, 181; air-cured, 12, 128; process defined, 116; color, 116-17, 137; quality of leaf, 116-17; high color, 117; paw-paws, 117; weather conditions, 118, 131, 137-38, 139; and barn design, 128; ventilation, 128, 130, 136-37, 138; fungus problems, 131; biological process of, 137; moisture content, 137, 138; stalk condition, 137; houseburn, 137-38; and split tobacco, 138; equipment, 138; burning coke, 138-39; anti-strut compounds, 139-40; strut, 139-40; sugar content, 180-81; technical description, 195 n 8.2. *See also* tobacco, conditions of
cutting, 116-28, **120**, **124**; timing, 115, 116, 124-25; dropping sticks, 117-18; splitting tobacco, 118-19, 119-20; and topping, 119; spearing tobacco, 120-24, **122**; setting sticks, 122-23; process of, 123, 125; splitting out, 123; pride in skill, 123-24; safety risks, 123-24; green tobacco sickness, 124; labor, 124; wilting, 124, 125-26; weather conditions, 125; cooping the tobacco, 126; handing up, 126; laying it down, 126; loading sticks, 126, **127**; harvesting machines, 127-28; stalk notching, 136. See also cutting and housing equipment; tobacco, conditions of
cutting and housing equipment, **120**; tobacco harvesting machines, 25-27; tobacco knife, 116; tobacco sticks, 116, 117, 118, 181; high-boy sprayer, 118; knife, shank or punch, 118-19; spear, 120, 121; tomahawk, 120, 120-21, 123; spear point, 121; tobacco wagons, 126; rail wagons, 126-27; harvesting machines, 127-28; tier rails, 131

dairying: alternative to tobacco, 1-2
Dark-fired Tobacco District Planters' Protective Association: Night Riders, 41
decision making: planting and marketing quota, 19-20; in landlord-tenant relationships, 35-36; tobacco variety, 74; planting by signs, 77; weather conditions, 84-85; uncertain future, 189. *See also* crop placement; crop rotation; grading; land use; marketing; packaging; timing; weather
delocalization: and agrochemicals, 185-87; transplants, 186
demography: average age of farmer, 3; out-migration, 30-31, 37; and labor supply, 188-89; older farmers, 188-89
disease, 14, 71, 102, 108; black root rot, 71-72; blue mold, 80, 109-10; black shank, 108-9. *See also* disease control; fungicides
disease control: nitrating the bed, 80; Ridomil, 80-81, 109-10; fungicides, 108-10; Nabam, 109, 195 n 7.1; disease resistance, 183. *See also* disease
Donahue, Burl: case, 143
Dooley, Hobert: tobacco sticks, manufacture of, 118
Duke, James B., 39-40, 43. *See also* American Tobacco Company
Duncan, George: tobacco labor and livestock, 26; shoe-type setter, 91-92; float plant system, 98-99; topping, 110; suckering, 112; harvesting machines, 128; baler, development of, 155; development of baling technology, 156-57; difficulty of mechanization, 180; barns and tobacco sticks, value of, 181; small family farm and mechanization, 181

economic importance: of burley tobacco in central Kentucky, 1-4, 11; decline in, 3, 190; to colonial U.S., 10
environmental tobacco smoke (ETS): and disease, 7. *See also* health problems; smoking
equipment: swapping, 29; care of, 81-82. *See also* cultivation equipment; cutting and housing equipment; field preparation equipment; setting equipment
erosion, 19

farmers' grades: flyings, 40-41, 147, 149, 188; red, 40-41, 147, 149; tips, 40-41, 47, 147; bright leaf, 47, 147; lugs, 47, 147, 149, 166, 188; red leaf,

Index

47; darker tip, 147; green, 147; light tan tip, 147; long bright, 147; long red, 147; long trash, 147; short bright, 147; short red, 147; trash, 147; leaf, 149, 188; long bright leaf, 166; brights, 188; cutters, 188
farmers' grades, poor quality: green, 159; rotten, 159; sunburned, 159; ground leaf, 159-60; dog bed tobacco, 160. *See also* stripping
farmers' marketing organizations, 189-90; Dark-fired Tobacco District Planters' Protective Association, 41; "tobacco wars" in central Kentucky, 41-46; Kentucky League of Tobacco growers, 42-43; American Society of Equity, 43; Burley Tobacco Growers Association, 43; Farmers and Tobacco Growers Association, 43; Henry County Union, 43; attempts to pool the crop, 43-45; Burley Tobacco Society, 43-46; Burley Tobacco Growers Cooperative Association, 47-50, 48. *See also* Burley Tobacco Growers Cooperative Association; Burley Tobacco Society; Night Riders
Farm Service Agency, 194. *See also* ASCS
FDA: as regulator of cigarettes, 192-93
fertility. *See* soil testing
fertilizer, 80-82; limestone, 16-17 (*see also* limestone); potash, 67; manure, 82; timing, 82; stalk as green manure, 160
fertilizer, chemical: early use, 14-15; ammonium nitrate, 20, 194 n 2.1; nitrogen, 20, 23-24; phosphorus, 20, 24; potassium, 20, 24; amounts used, 20-21; application methods, 20-24; economic benefits of, 21, 24; availability, 22; putting it in the row, 22; side dressing, 22; regulation of, 24; timing of applications, 24; and intensification, 182-83; soil acidity, 183
field preparation, 80-82, 83-86; plowing, 83-85; on hillsides, 84; soil moisture content, 84; timing, 84; dragging it down, 84-85; weather conditions, 84-85; laying off, 85-86. *See also* field preparation equipment

field preparation equipment: axe, 15; grubbing hoe, 15; jumping cutter plow, 15; drill, one-horse, 21; spreader, one-horse, 21; scoop for lime, 23; shovel, 23; spreader, bulk, 23; wagon, 23; plows, 83-85
fire-cured tobacco: USDA tobacco classification, 12
float plant system. *See* labor; seedling production; setting; tobacco seed
flue-cured tobacco, 180-81; priming, 12; USDA tobacco classification, 12
Fore, Joseph: origin of white burley tobacco, 11-12
Fowlkes, Charles: splitting the sale, 175-76
fungicides, 108-10; Ridomil, 80-81, 109-10; Nabam, 109, 195 n 7.1. *See also* disease control

garden crops: in tobacco beds, 77
Garner, Wightman: *The Production of Tobacco*, 195 n 8.2
Garrison, Clara: burning the beds, 66; sowing the beds, 75; cottons, care of, 77; cultivation, 106
government grades, 38, 146; leaf, 171; stalk position, 171; tips, 171; trashes, 171, 172; color, 171-72; quality, 171-72; flyings, 172; green, 172; lugs, 172; price, 172; variegated, 172. *See also* tobacco program
government grading, 171-72; government grader, 38, 146, 171-72; color, 171; group or leaf type, 171; ticket, 171; quality, 171-72. *See also* tobacco program
grading, 145-49, 185; effect of short production, 61; blending, 117; overripe, 117; buyers' grades, 146; farmers' grades, 146, 147-49, 149; stalk position, 146; use of leaf, 146; risk, 147; market price, 147-49; changing patterns of, 148; mixed grades, 148-49; top grade, 148-49; bales, 158; stick graded, 166; mixing grades, strategy, 173-74. *See also* marketing; packaging
Grady, Dan: sale records, 178
grain: as alternative crop, 1; in colonial trade, 11. *See also* crops, other
Great Depression, 189. *See also* New Deal, 51-54

Greene, Neva: economic importance of tobacco, 1-2; off-farm employment, women, 28; tobacco program, 54

Greene, Roy, 19; suitability of burley tobacco for central Kentucky, 13; crop placement, 19; fertilizer use, 24; yields with fertilizer use, 24; acreage-based allotment and overproduction, 56; field preparation, 84; housing, 126; stripping room, 142-43; high case, 144; grades, 146; farmers' grades, 147; changing patterns of grading, 148; baling and labor, 153; weight loss, 162; basket of hands, 165; government grading, 171-72; grading color and quality, 172; mixing grades, 173-74; tobacco auction schedule, 174; sales ribbons, 178

Haley, Kelly C.: crushing rock fences for lime, 23; lime applications, 23

hands, 165-66, 185; tying into hands, 149; head, 150, 165; tail, 150, 165; tie leaf, 150, 158, 195 n 9.1; pressing tobacco, 150-51, **151**; market price, 159; baskets, 165-66, **167, 168**; stick graded, 166; blending the tobacco, 173. *See also* marketing; packaging; stripping

hand-tied tobacco. *See* hands

Harney, Arthur Jr.: acreage-based allotment, 55-56; weeding the beds, 80; worming by hand, 108; suckering, 114; stripping, 147; changing patterns of grading, 148

harvesting tobacco. *See* cutting; housing

health problems, viii, 7-9; arteriosclerosis, 7; cancer, 7; chronic bronchitis, 7; emphysema, 7; hypertensive heart disease, 7; effects of smoking, 9; impact on tobacco culture, 9; risk of spraying for suckers, 115. *See also* cancer; nicotine; smoking

hemp: in colonial economy, 11

herbicides, 105-6, 185-86; timing, 82. *See also* cultivation; seedbed management; weeds

Hillenmeyer, Ernie: development of poundage system, 58

history of tobacco, colonial: and Native Americans, 3, 10; as historic icon, 3; in United States, 9-10; New World origins, 9-10; in Kentucky, 10-11; Spanish colonial presence, 10-11. *See also* Boonesborough, Ky.; Jamestown, Va.; Kentucky; Virginia

hogsheads, 185; early use of, in Boonesborough, Ky., 11; and storage of redried tobacco, 39; screwing header down, 41

Holland Transplanter Company: wheel-type setter, **92, 94**. *See also* setting equipment

housing, 116, 133-36; timing, 116, 124-25, 126; labor, 124; weather conditions, 124-25; wilting, 126; spacing of tobacco for curing, 131; barn man, 133; green tobacco, 133, 135; ground man, 133; ground squirrel, 133; handing off, 133, **134**, 135-36; hanger, 133; putting down tobacco, 133; spreaders, 133; wagon man, 133; hanging tobacco in the barn, 133-36; double-barning, 135; stick breakage, 135; climbing in the barn, 135-36; safety risks, 135-36. *See also* barn space; tobacco, conditions of

Immigration and Naturalization Service, 33

Immigration Reform and Control Act of 1986, 33

Imperial Tobacco Company, 39

insecticides, 185, 186

Ison, Louis: development of poundage system, 58

Jackson, Earl: cover crops, 18, 18-19; crop rotation, 18; soil testing, fertilizer use, 18

James I: *Counterblaste to Tobacco*, 194 n 1.1

Jamestown, Va.: economic importance of tobacco, 10; Native American agricultural practices, 10

Jones, Arthur D.: clearing, 15; crop rotation, 17-18; fertilizer application, 22; tenant subsistence operations, 34-35; weeds in tobacco beds, 67; field preparation, 84; tarring barns, 129-30; hauling tobacco, 163-64

Kautz, Fred: origin of white burley tobacco, 11-12

Index

Kennedy-Galbraith hand setter, **87,** 87-88
Kennedy-Galbraith Planter Company: Gateway tobacco press, 150-51
Kentucky: burley tobacco production, 1-4, 11; history of tobacco in, 10-12; dark-fired production region, 41; "tobacco wars" in central Kentucky, 41-46; Bluegrass region, 42. *See also* Boonesborough, Ky.; economic importance; farmers' marketing organizations; history of tobacco, colonial; Lexington, Ky.; University of Kentucky
Kentucky Farm Bureau, 58
Kentucky Migrant Network Coalition, 33. *See also* labor
Kentucky Tobacco Patch (Steed), x
Kiser, Eugene M.: economic importance of tobacco, 2; crop rotation, 17; labor, swapping, 29; suckering, 113
labor, 25-37, 102; cooperative, 25, 27, 29; family, 25, 27, 28, 188-89; hired, 25, 27, 29-33, 188-89; sharecropping, 25, 33-37; supply, 25, 29-33, 37, 179, 188-89; peak demand, 25-26; and livestock, 26; housing, 26; women, 27-28, 105, 136; quality of hired labor, 30; costs, 30-32, 31-32; and welfare, 31; migrant workers, 31, 32-33, 188-89, 194 n 3.2; division by sex, 74; float plant system, 98-99; and herbicide use, 105-6, 185; and spraying for suckers, 114, 115; custom work, 114; and newer barn design, 136; stripping, 148-49, 149; baling, 153-54, 157-58, 185; sheeting, 156; seasonal, in warehouses, 167; labor-saving innovations, 179; setting, 184; and MH sucker control, 185; use of agrochemicals, 185-86; African-American hired hands, 188-89; swapping, 188-89; local, 189. *See also* off-farm employment
labor supply: and demography, 188-89. *See also* labor
land use, 14-19; topography, 9, 16; fallowing, 14-15. *See also* crop placement; crop rotation
land values, 2, 58; and sale of marketing quota, 60-61
Lexington, Ky., 3, 43, 47, 60; Burley Tobacco Growers Cooperative Association, 38-39
Lexington Herald-Leader: market price for bales, 159
limestone: as soil nutrient, 16-17; and soil acidity (pH), 22; as fertilizer, 22-23; rock fences as source, 23
Little, Arthur: crop rotation, 19
livestock, 15, 16, 18; alternative to tobacco, 1-2; and labor, 26
Louderback, Spurgeon: setting by hand, 86; barn construction, 132-33
Louisville Courier-Journal: Bingham, Robert W., 47
Lugar, Senator Richard: proposed buyout of market quota, 193

managerial control: gassing the beds, 183-84; use of planting machines, 184; use of transplants, 184
Manley, Bill: stripping, 149
marketing, **168;** return on crop, 2, 40, 175; processing before sale, 5; in colonial Kentucky, 10-11; timing, 12, 162-63; quota, 19-20, 178; in keeping order, 40; sales in the barn, 40-41, 51; tobacco buyers, 40-41; tobacco receiving point, 41; WWI, effects of, 46; draw payments, 49; dump houses, 49-50; free-riders, 49-50, 50; pinhookers, 51; imported tobacco, 61, 62; short production, 61, 62, 148-49; and grading, 147-49; mixed grades, 148, 173-74; demand, 148-49, 173, 175; packaging, 149; handling, 151, 159, 166-67, 170, 175, 180, 187-88; price supports, 154; sheeting, 155-56; weight, 156, 162, 169; bales, 158-60, **164,** 166-67 (*see also* baling); dishonest packaging, 159, 161; poor quality leaf, 159-60; hogsheads, 161; loose-leaf auction system, 161; on the breaks, 161; Kentucky white burley market, 161-62; opening sale, 161-62, 165; warehouse calendar, 161-62; weight loss, 162; unloading tobacco, **163,** 164, **167;** hauling tobacco, 163-64; baskets, 164, 165-66, 167, 168; market price, 164-65, 170, 176-77; hand-tied tobacco, 165-66, 166 (*see also* hands); quality of leaf, 166; warehouse floor, 167; weighing, 167-68; marketing card, 168, 178; lots, 169; warehouse, 169-70; land

208 Index

marketing *(cont'd)*
lord-tenant economic arrangement, 170; tobacco on the floor, 170; government grading, 170-72; blending the tobacco, 173; price, 173; splitting the sale, 175-77; sales, 175-78; breaking the sale, 177; moving it off sale, 177; ribbons, 177; rejecting the sale, 177-78; tags, 178; changes, 180; settlement of 1997, 192-93. *See also* government grading; market price; marketing quota; packaging; tobacco auction; tobacco program; warehouse

marketing quota: allotment, 54; acreage-based allotment system, 55-57; overproduction, 56; poundage-based allotment system, 56; acreage reduction, 56-57; real estate values, 58; marketing card, 58-59; down in weight, 59; carry over, 59-60; lease prices, 60; leasing in poundage, 60; leasing out quota, 60; sale of tobacco quotas, 60-61; leasing for barn space, 135; switch from acre-based to poundage-based, 183. *See also* tobacco program

market price, 170, 176-77, 179; competition, absence of, 4, 4-5; seasonal timing, 4; information, sharing of, 4-5; impact of quality on, 5; attempts to control, prior to WWI, 41-46; farmers attempts to control, 41-51 (*see also* farmers' marketing organizations); effects of WWI on burley market, 46-47; price collapse of 1920, 47-48, **48;** parity price, concept of, 52; and grading, 187-88. *See also* marketing

Maryland tobacco, x

Massie, Ira: economic importance of tobacco in central Kentucky, 3-4, 4; crop rotation, 16; seed production and selection, on-farm, 70-71; varieties of tobacco, local, 71; overseeding the beds, 75; laying off, 85-86; season, 86; setting, time required, 88-89; cultivation with work animals, 103-4; hoeing by hand, 105; coking, 138; grading, risk, 147

Mechanical Transplanter Company: wheel-type setter, **92,** 94

mechanization: cutting and housing, 25-26; and migrant labor, 32; setting, 92-94, 100-101; cultivation, 102; stripping, 145-46; constraints to innovation, 180-81. *See also* change; equipment; technical innovations

migrant workers, 31, 32-33, 188-89, 194 n 3.2; Kentucky Migrant Network Coalition, 33. *See also* labor

Miller, Alex S.: mechanization of housing, 26; tenant farmers, 36; pest control, 107; topping, 111-12

Native American agricultural practices, 3, 10, 179

New Deal, 189, 190; Agricultural Adjustment Act, 51-54; programs, 22, 195 n 4.5. *See also* Agricultural Adjustment Act Committee

New Idea Spreader Co., Coldwater, Ohio: New Idea transplanter, **90**

New Orleans: as market for agricultural produce in colonial times, 11

Nicotiana tabacum, 9-10

nicotine: characteristics valued by users, 6; addiction, 6-7; and demand for tobacco products, 7; and health problems, 7-9; and public discourse, 9; cigarettes as delivery device, 192. *See also* cancer; health problems; smoking

Night Riders, **42;** Dark-fired Tobacco District Planters' Protective Association, 41; raids, 41, 44-45; Burley Tobacco Society, cut-out of 1908, 44-45

No-net-cost-program Act of 1982, 191; fees, 61, 62; tobacco program costs, 61. *See also* tobacco program

off-farm employment, 3, 25, 187, 188; women, 28; and labor supply, 29-30; and lease prices, 60

Olden, Kenneth: smoking, as cause of cancer, 8

O'Rourke, James: pegging it in, 86; cultivation with work animals, 104; hoeing by hand, 104-5; cutting, 119

out-migration, 30-31; impact on sharecropping, 37. *See also* demography; labor

Owens, Vivian: permit for MH use, 115

Index

packaging, 149-59; baling, 142, 149, 185; change, 149; hand-tied tobacco, 149, 185 (*see also* hands); sheeting, 149; rehandling, **151**; innovations in, 154-59; baskets, 165-66, **167, 168;** labor costs, 173; balers, 195 n 8.3. *See also* baling; hands; marketing; sheeting
Palmer, Gary: weather conditions, 106; soil acidity, 183
pest control: pesticides, 107, 108, 185; worming by hand, 107-8
pesticides, 107, 185; Paris Green, 108
pests, 102, 107
Philip Morris Tobacco Company: baler, development of, 155; buyers, 177
planting by signs, 77
plowing. *See* field preparation
pool. *See* Burley Tobacco Growers Cooperative Association; farmers' marketing organizations
population: out-migration, 30-31, 37. *See also* demography
price. *See* market price
pride: laying off, 86; in cutting skill, 123-24; packaging, 158; careful handling and display of crop, 170. *See also* aesthetics; tobacco culture
priming: flue-cured tobacco, 12; red burley, 12
processing: redriers, 39; redried tobacco, 54
profits, 2; and fertilizer use, 21. *See also* market price

quality, 159-60, 187; production practices, 5; and spraying for suckers, 115; of cured leaf, 116-17; and grading, 172; in good order, 180

Rankin, Bud: crop rotation, 19; fertilizer application methods, 22; New Idea Transplanter, 91; plant bed management, change in, 96; MH sucker control, 114; double-barning, 135; baling, 154-55; sheeting, 154-55
real estate values, 191
red burley tobacco, 11, 12. *See also* burley tobacco
Reese, Mark: migrant labor, 33
Reynolds, Eddie: suckering, 112-13
Richard, F.W.: pollination of hybrid tobacco, 74

Richards, Oscar: economic importance of tobacco, 2; soil tests, 23-24; renewal of tobacco program, 55; poundage system and carry over, 59; tobacco program, 62-63; planting by signs, 77; seedbed management, 79; field preparation, 85; weather conditions, 106-7; when to cut tobacco, 116-17; stripping, 149; baling, 153, 157-59; market price, 159; weight, 162; mixing grades, 173-74
risk: management, 102; reduction of, 183-84
R.J. Reynolds Tobacco Company: Camel brand cigarette, 13; use of burley tobacco, 13; buyers, 176-77
Robertson County Times Democrat, 22
Robinson, Lucian: economics, return on crop, 2; sowing the beds, 75, 76-77
Rolfe, John: tobacco cultivation as cash crop, 10
Roosevelt, Franklin D.: New Deal legislation, 51-52

season: setting, 86-87; in case for stripping, 143-44; for setting by hand, 184. *See also* case
seedbed. *See* seedbed management; seedbed preparation; setting
seedbed management, 79-81; placement, 64; protective covers, 64, 77-79, **78;** timing, 74; sowing the beds, 74-75; garden crops in bed, **77;** pulling plants, 79; weather conditions, 80; weeds, 80; off-season, 82
seedbed preparation, **76;** burning the beds, 15, 64-67, 183-84; use of debris from clearing, 15; fertilizer use, 64; soil preparation, 64-70; cutting wood for, 65, 66; dragging brush for, 65-67; timing, 66, 69; equipment, 67-69; steaming the beds, 67-69, 68; labor involved, 69; gassing the beds, 69-70, 183-84; custom work, 70
seedling production: float plant system, 64, **97,** 97-101; changes in, 83, 85, 100; seedlings as commodity, 97; use of transplants, 184
setting, 79, 80, 83-101; pulling the plants, **78,** 95-96, 98-99; spacing of plants, 83; timing, 83, 85, 88-89, 94; field preparation, 83-85; dropping the plants, 85,

setting *(cont'd)*
  88, **93**, 100-101; resetting, 85, 101; by hand, 85-89; pegging it in, 86, 184; weather conditions, 86-87, 94-95, 95, 101; use of hand setters, 87-89; missing the water, 90, 92, 94; with wheel-type setters, 92-94; healthy seedling, 95; clipping the plants, 95-96; with float plant system, 99; with one-row (wheel-type) setters, 100-101; and use of setter, 184. *See also* seedbed management; seedling production; setting equipment
setting equipment: hand setter, 15; drag, 65-67; sacks, 82, 195 n 5.1; setter, 83; peg, 85; Kennedy-Galbraith hand setter, **87**, 87-88; shoe-type setter, 89-92, **90;** New Idea transplanter, 90-91; Holland Transplanter Company wheel-type setter, **92;** wheel-type setter, 92-94, **93;** tobacco sticks, 181
sharecropping: landowner's reasons for, 33; economic arrangements, 33-34, 35; tenant subsistence operations, 34-35; decision-making in landlord-tenant relationships, 35-36
sheeting: development of, 154-55; description, 155; objections to, 155-56; testing, 156-57. *See also* marketing; packaging
Shuffett, Milton: labor and housing, 26; labor, supply of, 29-30; migrant labor, 32; market price and grading, 148-49
side dressing. *See* fertilizer
Simpson, Lawrence: wood cutting for seedbed preparation, 65; cutting, 125; coking, 139
Sims, Christine: Night Riders, 45; burning the beds, 65; garden crops in tobacco beds, 77; weeding the beds, 80; field preparation, 84; season for setting, 87
size of farm, 179-80, 186-87; consolidations, 191
Smiley, Joe: baler, development of, 155
smoking, 7-9; and ETS, 7; benzopyrene, 7-9; and health problems, 8-9, 194 n 1.1; and stress management, 9. *See also* cancer; nicotine
Snell, William: record quota, 193
Soil Conservation and Domestic Allotment Act of 1936, 53-54
soil fertility, 14-17, 18-24; fertilizer, 14-15; new ground, 15, 17; bottom land, 16; overflow ground, 16; hillsides, 16-17; limestone, 16-17; and crop rotation, 18; and fertilizer, 18, 20-24; and crop placement, 19; soil acidity (pH), 22, 183; and soil tests, 23-24. *See also* crop placement; crop rotation; fertilizer
soil testing, 23-24
sowing down the ground, 141; incorporating suckers, 113. *See also* cover crops
sowing the beds, 74-75. *See also* seedbed management
Spanish Embargo of New Orleans, 11. *See also* history of tobacco, colonial
Speedling Company: float plant system, 195 n 6.1
spraying for suckers: color, 114, 115; timing, 114; weather conditions, 114; fatty alcohols, 114-15; MH (maleic hydrazine), 114-15; health risk, 115; labor, 115; quality, 115; yields, 115. *See also* suckering
Steed, Virgil: *Kentucky Tobacco Patch,* x
Stokes, Granville: development of poundage system, 56, 57-58
strawberries, 2
stripping, 40, 142-60, **145;** putting it down, 133, 143-44, 158; timing, 140; lighting, 142; stripping room, 142, 145; case, 143-44, 158; weather conditions, 143-44; books, 144; bulk, 144, 149, 150, 152; hot tobacco, 144; stripping leaves off stalk by hand, 144-45; packaging, 145, 149-59; mechanized, 145-46; grading, 145-49, 147-48; stripped-out tobacco, 146; costs, 148-49; labor, 148-49; gum, 149; leaf fat, 149; process of, 149; stripping crew, 149; by hand, 149-51; stalks, 152; tobacco sticks, 152. *See also* case; farmers' grades; grading; hands; marketing; packaging; season; stripping room equipment
stripping room equipment, **145;** compressor, 141; hydraulic presses, 141; steam boiler, 141; baler, 142, 152, 153, 155, 157, 195 n 8.3; heater, 142; stick racks, 142, 149, 152; stripping

Index

table, 142, 149; tobacco sticks, 142, 149, 150; mechanical stripping machines, **145-46;** tobacco press, 150-51, **151;** wagon-bed presses, 151
Stull, Mike: stripping, 149
suckering, 5-6; spraying, 112, **113,** 114-15; suckers, 112; by hand, 112-14. *See also* spraying for suckers
suckering equipment: hooked-beak knife, 112; high-boy sprayer, **113,** 114; backpack sprayer, 114
swapping. *See* equipment; labor

technical innovations, 180; float plant system, 98-99; harvesting machines, 128; baling technology, 156-57; labor-saving, 179; constraints to innovation, 179-81; development and adoption, 191-92. *See also* change; mechanization
tenant farmers: family labor, 28. *See also* sharecropping
timing: market price, 4; marketing, 12, 162-63; fertilizer, 24, 82; seedbed preparation, 66, 69; seedbed, 74; herbicides, 82; setting, 83, 85, 88-89, 94; field preparation, 84; cultivation, 106; spraying for suckers, 114; cutting, 115, 116, 124-25; housing, 116, 124-25, 126; stripping, 140; topping, 112, 140
Toadvine, Evelyn: labor, 27; sharecropping, economic arrangements, 34
tobacco. *See* burley tobacco; farmers' marketing organizations; history of tobacco, colonial; tobacco culture; tobacco leaf; tobacco plant; tobacco production; tobacco program
tobacco, conditions of: mildew, 12; rotting, 12, 131; paw-paws, 117; sunburn, 119, 125-26; houseburn, 126, 131, 137-38, 144, 147; overcoked, 138; piebald, 138; strut, 139-40. *See also* cutting; housing; tobacco seedling
Tobacco and Kentucky (Axton), x
tobacco auction, 149, **174,** 174-78; history of, 161; auctioneer, 169, 175-76; lots, 169, 171; participants, 169; rows, 169; starter, 169, 175; ticket markers, 169; bidding, 171; rejecting the sale, 172; rotating schedule, 174; grade, 175; handling, 175; price, 175, 176-77; spotter, 175; starting price,

175; support price, 175; splitting the sale, 175-76; quotas, buyers', 176-77; allocation, 177; buyers, 177. *See also* government grading; marketing; tobacco program; warehouse
tobacco companies, 192-93; R.J. Reynolds, 13, 46, 176-77, 192; American Tobacco Company, 46; Liggett and Myers, 46; Lorillard, 46, 192; bidding, 147; blends, 147; quotas, 147; Southwestern, 177; B.A.T., 192; Brown and Williamson, 192; Philip Morris, 192; RJR Nabisco, 192; settlement of 1997, 192-93
tobacco culture, viii; commodity culture, 1-9; cultural distinctiveness, 3-4; cultural elaboration, 4; classification, 5; tobacco, botanical properties, 5-6, 180-81; impact of health consequences, 9, 179, 191; uncertain future, 179-80, 190, 191-93; changes, 179-93, 180, 182-91, 187-89; difference from flue-cured tobacco culture, 180-81; community welfare, 182; intensification, 182-83; decreased self-sufficiency, 185-87; changing categories, 187-88; changing demography and labor supply, 188-89; commodity in decline, 190; tobacco program, 190; symbolic devaluing of tobacco, 191-92. *See also* aesthetics; change; pride
Tobacco Inspection Act. *See* tobacco program
Tobacco Inspection Act of 1935: government inspection and grading, 53. *See also* government grading
tobacco leaf: fragility, 4, 5-6, 180; size of, 5-6; types, desirability of, 13; moisture content, 143-44, 195 n 10.1; poor quality, 159-60; handling, 180, 187-88. *See also* case; curing; farmers' grades; stripping
tobacco leaf type, 145-47, 159-60; tips, 145, 146, 147; brights, 146, 147; leaf, 146; long red, 146; lugs, 146; red, 146, 147; short red, 146; trashes, 146, 147; flyings, 147; green, 147
tobacco plant, 5-6; flowers, 5-6, 110; suckers, 5-6, 112; botanical properties, 5-6; healthy, 102; buttons, 110; tops, 110-12; leaf, 145-47; stalk, 160; disease resistance, 183. *See also*

tobacco plant *(cont'd)*
seedling; tobacco leaf type; tobacco seed
tobacco production: burley belt, 3; care and knowledge required, 4; fragile commodity, 4; and labor costs, 31-32; and Native Americans, 179; uncertainty in future, 179-80; constraints to innovation, 180; in good order, 180; technical innovations, 180; transformations, 180; investment in on-farm infrastructure, 181; intensification, 182-83; increases in managerial control, 183-84; risk reduction, 183-84; reduced costs, 185; use of agrochemicals, 185-86; decreased self-sufficiency, 185-87; quality tobacco, 187; handling, 187-88; demography and labor supply, 188-89; development and adoption of technology, 191-92. *See also* change; history of tobacco, colonial; Kentucky; labor; technical innovations; tobacco culture
tobacco program, 190, 191; administration, 38, 55 (see also ASCS); market price, stabilizing effect on, 4; marketing quota, 19-20, 54, 59, 62; role of warehouse, 38, 169; support prices, 38, 54, 54-55, 57, 62-63, 146, 171, 194 n 4.1; USDA, 38, 39; program costs, 38-39, 61; overproduction, 46-47, 56, 57-58; acreage-based quota system, 52-53, 55-57; poundage-based quota system, 52-53, 57-58; pool, 54-55, 61-62, 171; Farm Service Agency, 55; Farmers' Committee, 55; renewal of program, 55; Tobacco Workers' Conference, 57-58; and carryover of poundage, 59-60; fate of, 61; short production, 61; No-net-cost-program Act of 1982, 61-62, 191, 194 n 4.4; Tobacco Program Improvement Act of 1985, 62; government grades of leaf, 146; marketing, 164-65; marketing card, 168, 178; Burley Tobacco Program, 190; importance of, 190; uncertainty of future of, 190-91; potential results of abolition of, 191; reduction of price supports in 1985, 191; proposed buyout of market quota, 193. *See also* barn space;

Burley Tobacco Growers Cooperative Association; government grading; marketing quota
tobacco seed, 70-74, **76;** size of, 5-6; Trinidad, as original source, 10; seed processing, on-farm, 70-71; seed selection, 70-71; commercial seed production, 72-74, 100; F.W. Richards Seed Company, Winchester, Ky., 72-74; B.L. Kelley and Sons display ad, **73;** pelletized seed, 100; primed seed, 100. *See also* burley tobacco, varieties; seedling production
tobacco seedling: and soil conditions, 101, 106-7. *See also* setting
tobacco stalk: disease, 160; uses of, 160
tobacco variety. *See* burley tobacco
topography, 14; bottom land, 16; hillsides, 16-17; and crop placement, 19; and erosion, 19
topping, 110-12, **111;** tops, 110; timing, 112
transplanting. *See* setting
transplants. *See* seedling production
Trinidad: tobacco seed, original source of, 10

Uniroyal: MH, 114
University of Kentucky, 182; recommendations on use of animal manure on tobacco, 20; soil testing, 23; tobacco breeding programs, 71; recommendations for planting, 83; recommendations for clipping plants, 96; MH testing, 114
University of Kentucky Agricultural Cooperative Extension Service, x; and choice of tobacco variety, 74; prevention of blue mold, 109; barn plans, 130, 136; baler plans, 153; promotion of baling, 153-54
University of Kentucky College of Agriculture: Tobacco Workers' Conference, 57-58; development of baling technology, 156-57
University of Kentucky Experiment Station: fertilizer analysis, 24
USDA: administration of tobacco program, 38, 39; leasing regulations, stings, 60
USDA classification system for tobacco: cigar-binder, 12; cigar-filler, 12;

Index

cigar-wrapper, 12; fire-cured, 12; flue-cured, 12; air-cured, 12-13; miscellaneous, 13

U.S. Department of Agriculture: tobacco types, 128

values. *See* aesthetics; pride; tobacco culture

Varner, Willard: return on crops, 2; and fertilizer application, 21; yields with fertilizer use, 24; timing of housing, 124-25

Virginia: history of tobacco in, 10; and Kentucky, 10-11; Virginia Tidewater, 179. *See also* history of tobacco, colonial

warehouse, **168;** history of, 10-11; as independent cooperatives, 42-43; reform of practices, 53; scales, 53; weighmen, 53; role in administration of poundage system, 58-59; basket man, 165; sales charges, 165; solicitors, 165; baskets, 165-66; slipsheet, 166; pallet, 166-67; labor, 167; records, 167; warehouse floor, 167; and tobacco program, 169; fees, 169, 195 n 9.2; floor space, 169; loans, 169; lots, 169; relationship with growers, 169-70; highest price, 170; tobacco on the floor, 170. *See also* marketing; tobacco auction

weather, 184; seedbed, 80; decision making, 84-85; field preparation, 84-85; setting, 86-87, 94-95, 101; cultivation, 106-7; spraying for suckers, 114; housing, 124-25; cutting, 125; curing, 131, 137-39; stripping, 143-44. *See also* case; season; timing

Webb, George: origin of white burley tobacco, 11-12

weed control. *See* cultivation; herbicides; seedbed management; seedbed preparation; weeds

weeds, 80, 102; Johnson grass, 104-5

Wegner, Shirley: fertilizer application, 21-22; sharecropping, landlord-tenant relationships, 35-36; cultivation with work animals, 104; splitting tobacco, 119

weight, 156, 162, 169; down in weight, 59; weight loss, 162; weighing, 167-68. See also case; marketing quota

Wells, James D.: diseases of tobacco, 71-72; barns, 136

Whalen, Allen J.: soil tests, 23; women's work, 28; labor problems and tobacco production, 31-32; burning the beds, 65-66; field preparation, 84

Whaley, Sam: acreage-based allotment, 55

wheel-type setter: Kolk, Ray, 92-93; Poll, Ben, 92-93

white burley tobacco, 13; origins of, 11-12; advantages of, 12; description of plant, 12. *See also* air-cured tobacco; burley tobacco

Wilson, Russell E.: shift in cutting practices, 121

Wilson, William A.: crop placement, 17

Witt, Geneva: labor and women, 27-28

Witt, Nelson: return on crops, 2; and fertilizer application, 20-21; equipment, swapping, 29

women. *See* labor, women

worming, 107-8, 186

yields, 179; with fertilizer use, 21, 24; overproduction, 56-57; and seed selection, 71; and improved seed, 72; and weeds, 102; and spraying for suckers, 114-15; and intensification, 183